METHODS IN MOLECULAR BIOLOGY

Series Editor
John M. Walker
School of Life and Medical Sciences
University of Hertfordshire
Hatfield, Hertfordshire, UK

For further volumes:
http://www.springer.com/series/7651

For over 35 years, biological scientists have come to rely on the research protocols and methodologies in the critically acclaimed *Methods in Molecular Biology* series. The series was the first to introduce the step-by-step protocols approach that has become the standard in all biomedical protocol publishing. Each protocol is provided in readily-reproducible step-by-step fashion, opening with an introductory overview, a list of the materials and reagents needed to complete the experiment, and followed by a detailed procedure that is supported with a helpful notes section offering tips and tricks of the trade as well as troubleshooting advice. These hallmark features were introduced by series editor Dr. John Walker and constitute the key ingredient in each and every volume of the *Methods in Molecular Biology* series. Tested and trusted, comprehensive and reliable, all protocols from the series are indexed in PubMed.

Single Cell Metabolism

Methods and Protocols

Edited by

Bindesh Shrestha

Waters Corporation, Milford, MA, USA

Humana Press

Editor
Bindesh Shrestha
Waters Corporation
Milford, MA, USA

ISSN 1064-3745 ISSN 1940-6029 (electronic)
Methods in Molecular Biology
ISBN 978-1-4939-9829-6 ISBN 978-1-4939-9831-9 (eBook)
https://doi.org/10.1007/978-1-4939-9831-9

This Humana imprint is published by the registered company Springer Science+Business Media, LLC, part of Springer Nature.
The registered company address is: 233 Spring Street, New York, NY 10013, U.S.A.

Preface

It has taken a very long time to compile these chapters to a book form. I started this work when I was a research project director in a DARPA-funded project at the George Washington University and now finishing it during my scientific career at an instrument manufacturer.

First of all, I would like to thank all the authors who contributed to the project. Sincere heartfelt gratitude to those who waited a very long time to get these chapters published and heavy gratitude to those who submitted the remaining chapters quickly when some of the contributors pulled out at the last minute. Thank you Akos Vertes, Alan K. Jarmusch, Ales Svatos, Alfredo J. Ibáñez, Andrew M. Smith, Anthony Midey, Bernhard Spengler, Biswapriya B. Misra, Bo Yang, Christina R. Ferreira, Chunlai Tu, Clint M. Alfaro, Dhaka Ram Bhandari, Giulia Coliva, Guido F. Verbeck, Hernando Olivos, Jasmin Krismer, Jens Sobek, Jeremy L. Norris, John F. Cahill, Linda L. Allworth, Linwen Zhang, Mandy S. Phelps, Maria Fedorova, Martin Pabst, Ning Pan, Phuong Le, R. Graham Cooks, Renato Zenobi, Richard M. Caprioli, Robert F. Steinhoff, Rolf Brönnimann, Shweta Chitoor, Sichun Zhang, Sung Jun Lim, Sylwia A. Stopka, Tina Tsui, ValentinaPirro, Vilmos Kertesz, Wei Rao, Xiaochao Zhang, Xiaoyun Gong, Xingyu Si, Xinrong Zhang, Zhenwei Wei, and Zhibo Yang!

In addition to the authors, sincere gratitude to the relenting and supportive publishers and colleagues from Springer such as John Walker and Anna Rakovsky. It would be remiss not to mention constant encouragement from my parents and my spouse.

The chapters in this book showcase metabolite analysis of single cells. I had the pleasure of writing two bookend chapters of the book, the introduction in the beginning and the future direction at the end.Moreover, finally, thank you to readers who picked this book for your reference. I hope you find it useful.

Milford, MA, USA *Bindesh Shrestha*

Contents

Contributors

CLINT M. ALFARO • *Department of Chemistry and Center for Analytical Instrumentation Development, Purdue University, West Lafayette, IN, USA*

LINDA L. ALLWORTH • *Thomas Jefferson High School for Science and Technology, Alexandria, VA, USA*

DHAKA RAM BHANDARI • *Institute of Inorganic and Analytical Chemistry, Justus Liebig University Giessen, Giessen, Germany*

ROLF BRÖNNIMANN • *Swiss Federal Laboratories for Materials Science, Dübendorf, Switzerland*

JOHN F. CAHILL • *Oak Ridge National Laboratory, Oak Ridge, TN, USA*

RICHARD M. CAPRIOLI • *Mass Spectrometry Research Center, Department of Biochemistry, Vanderbilt University School of Medicine, Nashville, TN, USA*

SHWETA CHITOOR • *Department of Bioengineering, University of Illinois at Urbana-Champaign, Urbana, IL, USA; Micro and Nanotechnology Laboratory, University of Illinois at Urbana-Champaign, Urbana, IL, USA*

GIULIA COLIVA • *Faculty of Chemistry and Mineralogy, Institute of Bioanalytical Chemistry, University of Leipzig, Leipzig, Germany; Center for Biotechnology and Biomedicine, Universität Leipzig, Leipzig, Germany*

R. GRAHAM COOKS • *Department of Chemistry and Center for Analytical Instrumentation Development, Purdue University, West Lafayette, IN, USA*

MARIA FEDOROVA • *Faculty of Chemistry and Mineralogy, Institute of Bioanalytical Chemistry, University of Leipzig, Leipzig, Germany; Center for Biotechnology and Biomedicine, Universität Leipzig, Leipzig, Germany*

CHRISTINA R. FERREIRA • *Department of Chemistry and Center for Analytical Instrumentation Development, Purdue University, West Lafayette, IN, USA*

XIAOYUN GONG • *Department of Chemistry, Tsinghua University, Beijing, P.R. China*

ALFREDO J. IBÁÑEZ • *Department of Chemistry and Applied Biosciences, ETH Zurich, Zurich, Switzerland; Zurich PhD Program Molecular Life Sciences, Life Science Zurich, Zurich, Switzerland; Instituto de Ciencias Ómicas y Biotecnología Aplicada, Pontificia Universidad Católica del Perú, Lima, Peru*

ALAN K. JARMUSCH • *Department of Chemistry and Center for Analytical Instrumentation Development, Purdue University, West Lafayette, IN, USA*

VILMOS KERTESZ • *Oak Ridge National Laboratory, Oak Ridge, TN, USA*

JASMIN KRISMER • *Kantonsschule Limmata, Zurich, Switzerland; Department of Chemistry and Applied Biosciences, ETH Zurich, Zurich, Switzerland*

PHUONG LE • *Department of Bioengineering, University of Illinois at Urbana-Champaign, Urbana, IL, USA; Micro and Nanotechnology Laboratory, University of Illinois at Urbana-Champaign, Urbana, IL, USA*

SUNG JUN LIM • *Department of Bioengineering, University of Illinois at Urbana-Champaign, Urbana, IL, USA; Micro and Nanotechnology Laboratory, University of Illinois at Urbana-Champaign, Urbana, IL, USA*

ANTHONY MIDEY • *Waters Corporation, Milford, MA, USA*

BISWAPRIYA B. MISRA • *Center for Precision Medicine, Section of Molecular Medicine, Department of Internal Medicine, Wake Forest School of Medicine, Medical Center*

Boulevard, Winston-Salem, NC, USA; Department of Genetics, Texas Biomedical Research Institute, San Antonio, TX, USA

JEREMY L. NORRIS • *Mass Spectrometry Research Center, Department of Biochemistry, Vanderbilt University School of Medicine, Nashville, TN, USA*

HERNANDO OLIVOS • *Waters Corporation, Milford, MA, USA*

MARTIN PABST • *Department of Chemistry and Applied Biosciences, ETH Zurich, Zurich, Switzerland; Department of Biotechnology, TU Delft, Delft, The Netherlands*

NING PAN • *Department of Chemistry and Biochemistry, University of Oklahoma, Norman, OK, USA*

MANDY S. PHELPS • *University of North Texas, Denton, TX, USA*

VALENTINA PIRRO • *Department of Chemistry and Center for Analytical Instrumentation Development, Purdue University, West Lafayette, IN, USA*

WEI RAO • *Department of Chemistry and Biochemistry, University of Oklahoma, Norman, OK, USA*

BINDESH SHRESTHA • *Waters Corporation, Milford, MA, USA*

XINGYU SI • *Department of Chemistry, Tsinghua University, Beijing, P.R. China*

ANDREW M. SMITH • *Department of Bioengineering, University of Illinois at Urbana-Champaign, Urbana, IL, USA; Micro and Nanotechnology Laboratory, University of Illinois at Urbana-Champaign, Urbana, IL, USA; Department of Materials Science and Engineering, University of Illinois at Urbana-Champaign, Urbana, IL, USA*

JENS SOBEK • *Functional Genomics Center Zürich, ETH Zurich and University of Zurich, Zurich, Switzerland*

BERNHARD SPENGLER • *Institute of Inorganic and Analytical Chemistry, Justus Liebig University Giessen, Giessen, Germany*

ROBERT F. STEINHOFF • *Novartis Pharma AG, Stein, Switzerland; Department of Chemistry and Applied Biosciences, ETH Zurich, Zurich, Switzerland*

SYLWIA A. STOPKA • *Department of Chemistry, The George Washington University, Washington, DC, USA*

ALES SVATOS • *Research Group Mass Spectrometry, Max Planck Institute for Chemical Ecology, Jena, Germany*

TINA TSUI • *Mass Spectrometry Research Center, Department of Biochemistry, Vanderbilt University School of Medicine, Nashville, TN, USA*

CHUNLAI TU • *Department of Bioengineering, University of Illinois at Urbana-Champaign, Urbana, IL, USA; Micro and Nanotechnology Laboratory, University of Illinois at Urbana-Champaign, Urbana, IL, USA*

GUIDO F. VERBECK • *University of North Texas, Denton, TX, USA*

AKOS VERTES • *Department of Chemistry, The George Washington University, Washington, DC, USA*

ZHENWEI WEI • *Department of Chemistry, Tsinghua University, Beijing, P.R. China*

BO YANG • *Mass Spectrometry Research Center, Department of Biochemistry, Vanderbilt University School of Medicine, Nashville, TN, USA*

ZHIBO YANG • *Department of Chemistry and Biochemistry, University of Oklahoma, Norman, OK, USA*

RENATO ZENOBI • *Department of Chemistry and Applied Biosciences, ETH Zurich, Zurich, Switzerland*

Linwen Zhang • *Biomolecular Measurement Division, National Institute of Standards and Technology, Gaithersburg, MB, USA; Institute for Bioscience and Biotechnology Research, Rockville, MB, USA*
Sichun Zhang • *Department of Chemistry, Tsinghua University, Beijing, P.R. China*
Xiaochao Zhang • *Department of Chemistry, Tsinghua University, Beijing, P.R. China*
Xinrong Zhang • *Department of Chemistry, Tsinghua University, Beijing, P.R. China*

Chapter 1

Single-Cell Metabolomics by Mass Spectrometry

Bindesh Shrestha

Abstract

Single-cell level metabolomics gives a snapshot of small molecules, intermediates, and products of cellular metabolism within a biological system. These small molecules, typically less than 1 kDa in molecular weight, often provide the basis of biochemical heterogeneity within cells. The molecular differences between cells with a cell type are often attributed to random stochastic biochemical processes, cell cycle stages, environmental stress, and diseased states. In this chapter, current limitations and challenges in single-cell analysis by mass spectrometry will be discussed alongside the prospects of single-cell metabolomics in systems biology. A few selected example of the recent development in mass spectrometry tools to unravel single-cell metabolomics will be described as well.

Key words Single-cell analysis, Single cell, Single-cell mass spectrometry, Single-cell metabolomics, Single-cell metabolites

1 Introduction

A single cell is a natural unit of life. The single-cell metabolomics analysis provides molecular information of small phenotypically relevant molecules without inferences made by averages of the heterogeneity of bulk tissue. Multiomics analysis of molecules in single cells can provide insights into the cell's phenotype, diseased states, and environmental influences [1]. Metabolomics analysis of single-cell offers an insight into the functional process within a cell. Dysregulation of one of these single cells in multicellular organisms, such as humans, can lead to diseases such as cancers and neurological disorders.

Analysis of single-cell structures can be done with a host of analytical technologies such as nanoscale, stimulated emission depletion (STED) microscopy, stochastic optical reconstruction microscopy (STORM), photoactivated localization microscopy (PALM), near-field scanning optical microscopy (NSOM), and scanning ion conductance microscopy (SICM) [2, 3]. However, no other technique comes close to the analytical merit of mass

Bindesh Shrestha (ed.), *Single Cell Metabolism: Methods and Protocols*, Methods in Molecular Biology, vol. 2064, https://doi.org/10.1007/978-1-4939-9831-9_1, © Springer Science+Business Media, LLC, part of Springer Nature 2020

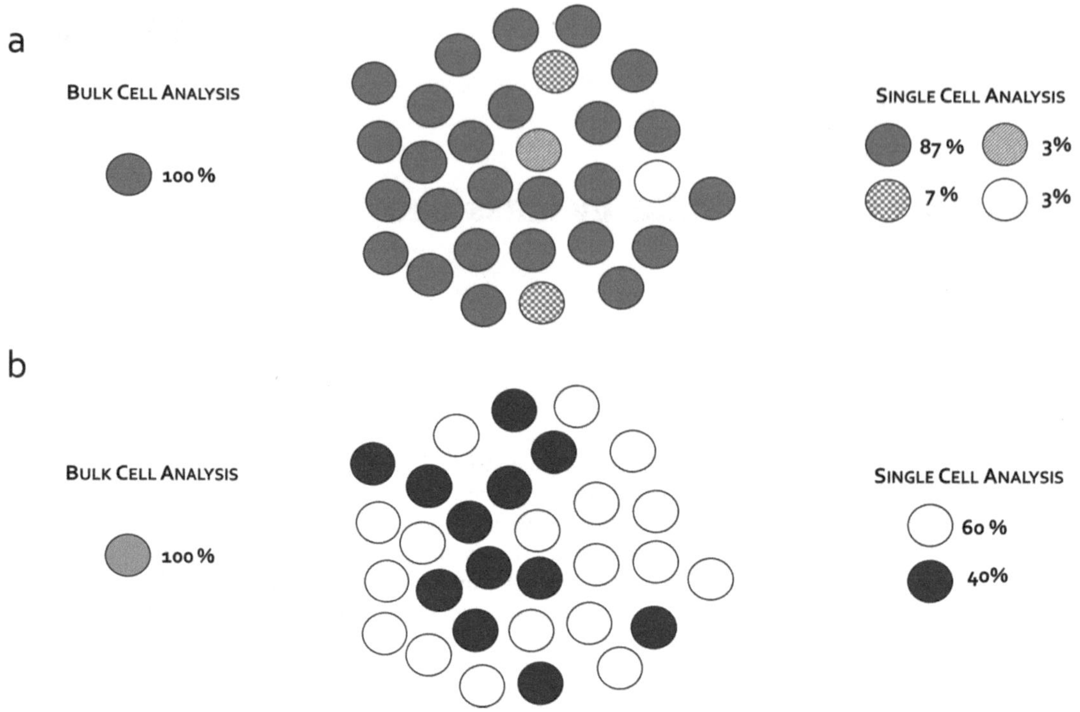

Fig. 1 Difference between bulk-cell analysis (**a**) vs. single-cell analysis (**b**)

spectrometry (MS) in selectivity and multiplexing for unlabeled single-cell analysis of metabolites. Due to enhancements in the hardware of mass spectrometer technologies, MS-based methods are reaching the sensitivity level where molecular profiling from a single cell can be detected within the heterogeneous cell population. The interest in single-cell mass spectrometry research is growing as shown by increased counts of academic publications with phrases "single cell" and "mass spectrometry" phrases. A conceptual difference between single-cell analysis and bulk cell analysis is shown in Fig. 1. The bulk analysis of cell population predominated with one type of cells will show all cells having the same features ignoring rare cells as shown in Fig. 1a. In contrast, cells with a bimodal distribution of two types of cells may show result indicating the average of two cell types as shown in Fig. 1b. Only single-cell analysis is capable of showing the accurate analytical characteristic of such a cell population.

Broadly, there are three categories of analytical workflow available to analyze cellular metabolism. The first one is to monitor the general input and output of metabolism, such as oxygen consumption rates, to infer metabolic activity. The second is metabolic enzymes characterization using enzyme activity assays to study metabolite activities. Finally, the third one is to utilize steady-state metabolomics analysis using mass spectrometry. The first method has not been used to characterize metabolites of single cell within their native microenvironment. The second method has shown some promise by using visual quantification of multiple enzymatic activities measurement combined with cell type identification that can be used to assess the metabolic configuration of single cells within their native tissue microenvironment [4]. Traditional biochemical method grinds up pellets of cells and analyzing the content giving an average molecular profile. Understanding cellular heterogeneity is vital to studying biological processes, such as the one involved in cancer and development. However, metabolite heterogeneity within the same cell population is largely unmapped because of the limitations of existing analytical tools for reliably quantifying metabolite levels in single cells. Analytical volume in single cells can range from 1 femtoliter (10^{-15}) in *Escherichia coli* to 1 picoliter (10^{-12}) in typical *Homo Sapiens* cells.

Currently, single-cell MS can be separated into two categories analysis, that of embedded dead cells or free-flowing live cells. Analysis of live cells is a much more delicate process, mostly using nano-ESI capillaries to extract content from a cell before using the same capillary to inject the contents into the MS directly [5–7]. Conceptual schematics of workflow for analyzing single cells by MS are shown in Fig. 2—single cells can be analyzed by desorption of cells, extracting the content of single cells or sorting and ionizing cells. The biggest challenges in single-cell metabolism analysis by mass spectrometry are the low amount of metabolites in individual cells, complexity in extracting small volumes of the sample for analysis from single cell, the wide dynamic range in concentration of metabolites present within a cell, quick turnover of metabolites, and the diversity of molecular species for metabolites. Mass spectrometry (MS) is a promising tool for both intra/extracellular metabolite analyses of single cells due to its high specificity, multiplexing capability, adequate sensitivity, and able to perform analysis without a need of molecular chaperone such as labeling. Strategies for sample treatment, separation methods, and data analysis require special considerations for single cells. Ongoing analytical challenges include subcellular heterogeneity, non-normal statistical distributions of cellular properties, and the need for high-throughput, high molecular coverage, and minimal perturbation. Some recent reviews referenced here have done an excellent job at reviewing the current status of mass spectrometry-based single-cell analysis [8–12].

Sorting or sampling of single cells is required before the single-cell analysis. The sorting of single cells depends on access to the type of cells, namely adherent, tissue embedded, and suspended cells. Cells are sorted mostly based on differences in cell size, morphology, and protein expression. The resulting homogenous populations of cells have essential applications in research and as therapeutics. Flow cytometry is used to analyze cell suspension by

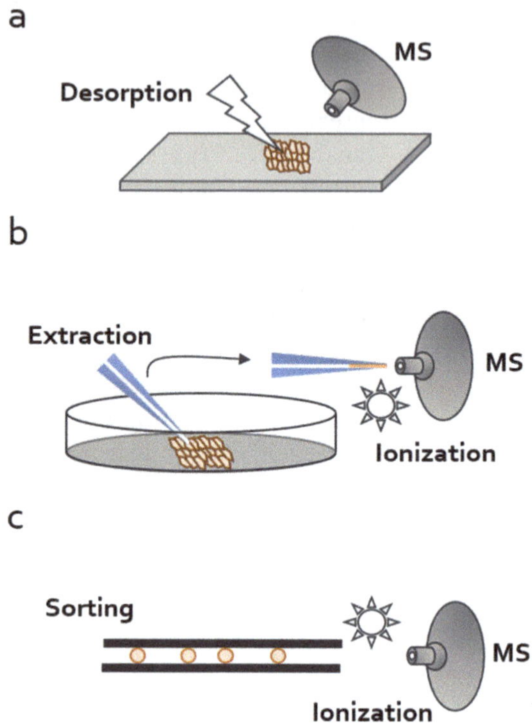

Fig. 2 Various common conceptual workflows for analyzing single cells by a mass spectrometer are presented

flowing a single cell at a time. Single cells are usually labeled with fluorescent markers, so when a laser beam is scattered through them, it can shed light into various cells properties. Mass cytometry, which combines flow cytometry and MS, has been used for sorting and targeted high-throughput molecular analysis of the suspended mammalian single cells [13]. In mass cytometry, antibodies are labeled with heavy metal ion tags for detection by inductively coupled plasma (ICP)-MS. [14] Mass cytometry system, such as CyTOF, ionizes the metal tags in free-flowing single cells by ICP followed by their detection by a time-of-flight (ToF). The ion intensities of detected metal ions are taken as surrogates for the tagged analyte. Mass cytometry is currently limited to a couple of dozen available proteins and not shown for metabolite analysis. Sorting and selection of single cells or subcellular component can be done by laser capture microdissection [15]. Sorting and analyzing cells using microfluidic chip has been demonstrated for mass spectrometry workflows. In laser desorption workflow microfluidic platform is used for off-line deposition of samples; while in electrospray-based platforms, the metabolic analysis of cells could be detected in offline or online modalities [16].

Most of the single-cell sampling for MS analysis was done by a desorption event, such as laser-based irradiance. The desorption-based MS techniques have been utilized to analyze metabolites within a single cell such as secondary ion MS (SIMS), matrix-assisted laser desorption/ionization (MALDI) [17], laser capture microdissection MALDI [18], matrix-free laser desorption ionization laser ionization from silicon nanopost arrays (NAPA) [19], nanostructure-initiator mass spectrometry (NIMS) [19, 20], laser ablation electrospray ionization (LAESI) [6, 21], and desorption/ionization on porous silicon (DIOS). Other MS techniques include utilizing cell micro-sampling event using micromanipulator driven needles and extraction of a single cell by using a probe-based micro-sampling event followed by electrospray (ESI) [7, 22–25].

Mass spectrometry technique does not have amplification process, such as polymerase chain reaction (PCR) to detect low abundant transcripts at a single-cell level. Unlike DNA and RNA, small molecules such as metabolites and lipids cannot be amplified. However, separation techniques combined with mass spectrometry can provide signal-to-noise ratio enhancement of low abundance ions by mitigating effects of matrix interference and minimize ion suppression before detection by MS. A single cell contains at least tens of thousands of molecular species. Direct analysis of such complex sample may be affected by ion interference and ion suppression. A separation strategy can reduce the complexity of MS analysis and enhance molecular coverage. Gas-based separation strategy, such as ion mobility separation (IMS), has shown to provide rapid separation for volume-limited single cells sample inaccessible by traditional liquid chromatography separation (LC/MS), often default choice for hyphenated separation technique for mass spectrometry. IMS is capable of separating isobaric ions within milliseconds after ionization and thus a promising choice in combination with desorption-based single-cell MS and mass spectrometry imaging [26]. As an additional benefit, collisional cross sections (CCS) of molecules can also provide an orthogonal identifier of ions by providing information on their size [27–29]. Other separation techniques used for single-cell metabolite analysis is capillary electrophoresis (CE)-ESI, which have been used to metabolic analysis in Xenopus laevis embryo [30, 31]. Liquid chromatography (LC) has been exploited to perform only single-cell protein analysis with tandem mass tags (TMT). In Single Cell ProtEomics by Mass Spectrometry (SCoPE-MS), single cells are manually picked under a microscope and mechanically lysed and followed by application TMT of single-cell augmented with about 200 carrier cells to obtain an adequate signal for peptide sequence [32]. In the future, we can anticipate a similar tagging workflow to explore single-cell metabolomics.

In the following chapters, learn more about single-cell analysis by micro-sampling coupled with electrospray in Chapters 2–4, electrospray probe designed for single-cell analysis in Chapter 5, laser capture microdissection coupled with liquid vortex capture in

Chapter 6, microarrays for MS (MAMS) in Chapter 7, MALDI analysis of single cells from Chapters 8–10, another laser desorption method NAPA in Chapter 11, quantum dots for cytology in Chapter 12, and desorption electrospray ionization (DESI) future in single-cell analysis in Chapters 13 and 14, open-source software tools and database in Chapter 15, and future directions in the final Chapter 16.

Methods for single-cell metabolism are still in their infancy and several fundamental challenges remain to move the technology from the hand of few experts to practicing biomedical scientist. Some of the significant challenges are outlined in the last chapter of the book, such as broader coverage of metabolites, accurate identification of metabolites, quantitative measurement, etc.

2 Conclusion

Single-cell metabolism analysis enables the accurate biochemical characterization of a cell within heterogeneous cell populations providing insight into cellular mechanism. Metabolic reprogramming can drive cell differentiation and its fate, cascading into the whole organism. Over the past years, single-cell metabolism has many aspects of cancer. New analytical tools will be required to extract metabolic information from single cells. There are several ways of analyzing single cells by mass spectrometry from laser capture microdissection to simply using an electrospray emitter needle to extract single-cell contents. Single-cell MS is an emerging field, and the current methods are still limited to the detection of either hundreds of metabolites from thousands of cells or detection of a handful of small molecule, most abundant ones, from an individual cell. Ongoing technological advances will see improvements in coverage and sensitivity. In the future, we can anticipate the continuous development of the methods and tools to address these challenges that will enable us to decipher the metabolic signature of complex and heterogeneous cell populations. Ongoing technological advances will lead to enhanced metabolite coverage and sensitivity, thus increasing the number of metabolites that can be studied simultaneously.

References

1. Macaulay IC, Ponting CP, Voet T (2017) Single-cell multiomics: multiple measurements from single cells. Trends Genet 33(2):155–168
2. Zheng XT, Li CM (2012) Single cell analysis at the nanoscale. Chem Soc Rev 41 (6):2061–2071
3. Galler K, Bräutigam K, Große C, Popp JR, Neugebauer U (2014) Making a big thing of a small cell ? Recent advances in single cell analysis. Analyst 139(6):1237–1273
4. Miller A, Nagy C, Knapp B, Laengle J, Ponweiser E, Groeger M, Starkl P, Bergmann M, Wagner O, Haschemi A (2017)

Exploring metabolic configurations of single cells within complex tissue microenvironments. Cell Metab 26(5):788–800.e6

5. Fujii T, Matsuda S, Tejedor ML, Esaki T, Sakane I, Mizuno H, Tsuyama N, Masujima T (2015) Direct metabolomics for plant cells by live single-cell mass spectrometry. Nat Protoc 10(9):1445
6. Shrestha B, Vertes A (2009) In situ metabolic profiling of single cells by laser ablation electrospray ionization mass spectrometry. Anal Chem 81(20):8265–8271
7. Zhang L, Foreman DP, Grant PA, Shrestha B, Moody SA, Villiers F, Kwak JM, Vertes A (2014) In situ metabolic analysis of single plant cells by capillary microsampling and electrospray ionization mass spectrometry with ion mobility separation. Analyst 139 (20):5079–5085
8. Heath JR, Ribas A, Mischel PS (2016) Single-cell analysis tools for drug discovery and development. Nat Rev Drug Discov 15(3):204
9. Trouillon R, Passarelli MK, Wang J, Kurczy ME, Ewing AG (2013) Chemical analysis of single cells. Anal Chem 85(2):522–542
10. Zenobi R (2013) Single-cell metabolomics: analytical and biological perspectives. Science 342(6163):1243259
11. Zhang L, Vertes A (2018) Single-cell mass spectrometry approaches to explore cellular heterogeneity. Angew Chem Int Ed 57 (17):4466–4477
12. Wang D, Bodovitz S (2010) Single cell analysis: the new frontier in 'omics'. Trends Biotechnol 28(6):281–290
13. Yao Y, Liu R, Shin MS, Trentalange M, Allore H, Nassar A, Kang I, Pober JS, Montgomery RR (2014) CyTOF supports efficient detection of immune cell subsets from small samples. J Immunol Methods 415:1–5
14. Bandura DR, Baranov VI, Ornatsky OI, Antonov A, Kinach R, Lou X, Pavlov S, Vorobiev S, Dick JE, Tanner SD (2009) Mass cytometry: technique for real time single cell multitarget immunoassay based on inductively coupled plasma time-of-flight mass spectrometry. Anal Chem 81(16):6813–6822
15. Cahill JF, Kertesz V, Van Berkel GJ (2016) Laser dissection sampling modes for direct mass spectral analysis. Rapid Commun Mass Spectrom 30(5):611–619
16. Mao S, Li W, Zhang Q, Zhang W, Huang Q, Lin J-M (2018) Cell analysis on chip-mass spectrometry. TrAC Trend Anal Chem 107:43–59
17. Li L, Garden RW, Sweedler JV (2000) Single-cell MALDI: a new tool for direct peptide profiling. Trends Biotechnol 18(4):151–160
18. Bhattacharya SH, Gal AA, Murray KK (2003) Laser capture microdissection MALDI for direct analysis of archival tissue. J Proteome Res 2(1):95–98
19. Walker BN, Stolee JA, Vertes A (2012) Nanophotonic ionization for ultratrace and single-cell analysis by mass spectrometry. Anal Chem 84(18):7756–7762
20. Northen TR, Yanes O, Northen MT, Marrinucci D, Uritboonthai W, Apon J, Golledge SL, Nordstrom A, Siuzdak G (2007) Clathrate nanostructures for mass spectrometry. Nature 449(7165):1033–1036
21. Shrestha B, Sripadi P, Reschke BR, Henderson HD, Powell MJ, Moody SA, Vertes A (2014) Subcellular metabolite and lipid analysis of Xenopus laevis eggs by LAESI mass spectrometry. PLoS One 9(12):e115173
22. Mizuno H, Tsuyama N, Harada T, Masujima T (2008) Live single-cell video-mass spectrometry for cellular and subcellular molecular detection and cell classification. J Mass Spectrom 43 (12):1692–1700
23. Tsuyama N, Mizuno H, Tokunaga E, Masujima T (2008) Live single-cell molecular analysis by video-mass spectrometry. Anal Sci 24 (5):559–561
24. Gong X, Zhao Y, Cai S, Fu S, Yang C, Zhang S, Zhang X (2014) Single cell analysis with probe ESI-mass spectrometry: detection of metabolites at cellular and subcellular levels. Anal Chem 86(8):3809–3816
25. Zhang L, Vertes A (2015) Energy charge, redox state, and metabolite turnover in single human hepatocytes revealed by capillary microsampling mass spectrometry. Anal Chem 87 (20):10397–10405
26. Stopka SA, Agtuca BJ, Koppenaal DW, Paša-Tolić L, Stacey G, Vertes A, Anderton CR (2017) Laser-ablation electrospray ionization mass spectrometry with ion mobility separation reveals metabolites in the symbiotic interactions of soybean roots and rhizobia. Plant J 91(2):340–354
27. Zhang X, Quinn K, Cruickshank-Quinn C, Reisdorph R, Reisdorph N (2018) The application of ion mobility mass spectrometry to metabolomics. Curr Opin Chem Biol 42:60–66
28. Paglia G, Williams JP, Menikarachchi L, Thompson JW, Tyldesley-Worster R, Halldórsson S d, Rolfsson O, Moseley A, Grant D, Langridge J (2014) Ion mobility derived

collision cross sections to support metabolomics applications. Anal Chem 86 (8):3985–3993

29. Zhou Z, Tu J, Zhu Z-J (2018) Advancing the large-scale CCS database for metabolomics and lipidomics at the machine-learning era. Curr Opin Chem Biol 42:34–41
30. Onjiko RM, Portero EP, Moody SA, Nemes P (2017) In situ microprobe single-cell capillary electrophoresis mass spectrometry: metabolic reorganization in single differentiating cells in the live vertebrate (Xenopus laevis) embryo. Anal Chem 89(13):7069–7076
31. Portero EP, Nemes P (2019) Dual cationic–anionic profiling of metabolites in a single identified cell in a live Xenopus laevis embryo by microprobe CE-ESI-MS. Analyst 144 (3):892–900
32. Budnik B, Levy E, Harmange G, Slavov N (2018) SCoPE-MS: mass spectrometry of single mammalian cells quantifies proteome heterogeneity during cell differentiation. Genome Biol 19(1):161

Chapter 2

Identification of Metabolites in Single Cells by Ion Mobility Separation and Mass Spectrometry

Linwen Zhang, Linda L. Allworth, and Akos Vertes

Abstract

Non-targeted metabolic analysis of single cells by mass spectrometry (MS) is important for understanding individual cell functions and characterizing cell-to-cell heterogeneity. However, identifying biomolecules in single cells presents significant challenges due to the low picoliter volume samples and the structural diversity of metabolites. Capillary microsampling electrospray ionization (ESI) MS with ion mobility separation (IMS) enables the analysis of single cells under ambient conditions with minimum sample pretreatment and improved specificity. Here, we describe a protocol for the analysis of the metabolic makeup, and the identification of ions produced from single cells by capillary microsampling ESI-IMS-MS.

Key words Single-cell analysis, Metabolomics, Mass spectrometry, Ion mobility separation, Isobaric ions, Collision cross-section

1 Introduction

Metabolic analysis of individual cells within an isogenic population enables the characterization of cellular physiological states and their phenotypic heterogeneity [1–3]. With high sensitivity and specificity, mass spectrometry (MS) is a valuable tool for the non-targeted analysis of a wide range of metabolites and lipids from single cells [4, 5]. High-performance liquid chromatography (HPLC) combined with MS is a routine method for the analysis of complex biological samples with high specificity [6, 7]. However, this method requires elaborate sample pretreatment and is not compatible with the metabolic analysis of single human cells with low picoliter volumes. Ion mobility separation (IMS) integrated with MS can be used to distinguish isobaric ions in the gas phase, based on their size, shape, charge, and mass, on a timescale of milliseconds [8–10]. The collision cross-section (CCS) values measured by IMS can enhance metabolite assignment confidence [11, 12]. In addition, the combination of collision-induced dissociation (CID)

Bindesh Shrestha (ed.), *Single Cell Metabolism: Methods and Protocols*, Methods in Molecular Biology, vol. 2064, https://doi.org/10.1007/978-1-4939-9831-9_2,

and IMS can improve specificity for the distinction and identification of isobars, isomers, and conformers [13].

Capillary microsampling and electrospray ionization (ESI) MS with IMS has been demonstrated for the metabolic analysis of single plant cells with improved molecular coverage [14]. This technique has been extended to subpopulations of human cells in particular stages of mitosis, and to the subcellular analysis of peptides in single snail neurons with identified function [15–17]. Here, we describe a protocol for analyzing small metabolites and lipids, and distinguishing isobaric ions in single adherent mammalian cells. Picoliter volumes of subcellular contents can be sampled and analyzed by this technique. The protocol does not require the detachment of the cells from the substrate surface or tissue for analysis, thereby avoiding the perturbation of metabolic state associated with scraping or trypsinization.

2 Materials

2.1 Reagents and Chemicals

1. Electrospray solution: HPLC grade methanol:water (*v/v* 4:1) with 1 mM ammonium formate (HPLC grade, Sigma-Aldrich, St. Louis, MO, USA).
2. For CCS determination, poly-DL-alanine (P9003, Sigma-Aldrich, St. Louis, MO, USA) was used as the calibrant and dissolved in the electrospray solution to reach a final concentration of 0.1 g/L.
3. One pouch of phosphate buffered saline (PBS) (Bioperformance certified grade, Sigma-Aldrich, St. Louis, MO, USA) was dissolved in 1.0 L deionized water (18.2 MΩ·cm) to prepare 1× PBS solution.
4. 2-Deoxy-D-glucose (2-DG) (≥99.0% purity, Sigma-Aldrich, St. Louis, MO, USA) stock solution was prepared in 1× PBS solution to reach a final concentration of 500 mM.

2.2 Cell Culture

1. Adherent mammalian cell lines, such as HepG2/C3A (CRL-10741, ATCC, Manassas, VA, USA).
2. Cell culture medium for HepG2/C3A: Eagle's Minimum Essential Medium (ATCC, Manassas, VA, USA) supplemented with 10% (*v/v*) fetal bovine serum and 1% (*v/v*) penicillin-streptomycin (Invitrogen, Grand Island, NY, USA).
3. An automated cell counter (Countess, Invitrogen, Grand Island, NY, USA) was used for determining the viable cell numbers using trypan blue.
4. 2 mL of cell suspensions at a density of 2×10^5 cells/mL was seeded on a 35 mm Petri dish (Corning, Tewksbury, MA, USA). Cells were maintained at 37 °C and 5% CO_2 in an incubator (HERAcell 150i, Thermo Scientific, Waltham, MA, USA) for 12–16 h before MS analysis (*see* **Note 1**).

2.3 Capillary Fabrication

1. A micropipette puller (P-1000, Sutter Instrument, Novato, CA, USA) installed with a box filament (FB255B, Sutter Instrument, Novato, CA, USA) was used for capillary fabrication.
2. Thin-walled borosilicate glass capillary with a filament (TW100F-3, World Precision Instruments, Sarasota, FL, USA) was chosen to produce relatively large tip openings (~1 μm) for adherent mammalian cell sampling. The filament in the capillary is necessary for drawing the electrospray solution to the end of the tip and minimizing air bubbles in the sample (*see* **Note 2**).
3. Pipette storage box (Sutter Instrument, Novato, CA, USA).

2.4 Cell Sampling

1. An inverted microscope (IX71, Olympus, Tokyo, Japan) with a maximum magnification of 600× was used for the visualization of cells.
2. A micromanipulator (TransferMan NK2, Eppendorf, Hauppauge, NY, USA) was mounted on the microscope for performing cell sampling.
3. A capillary holder (IM-H1, Narishige, Tokyo, Japan) was attached with a syringe.

2.5 Single-Cell ESI-IMS-MS

1. Microloader tips (Cat No. 930001007, Eppendorf, Hauppauge, NY, USA) for backfilling the capillary with the electrospray solution.
2. A microelectrode holder (MEW-F10A, Warner Instruments, Hamden, CT, USA) with a platinum wire of 200 μm in diameter and ~5 cm in length (Alfa Aesar, Ward Hill, MA, USA) provided electrical connection to the electrospray solution in the capillary tip. High voltage was applied to the microelectrode by a regulated power supply (PS350, Stanford Research Systems Inc., Sunnyvale, CA, USA) to generate an electrospray (*see* **Note 3**).
3. A quadrupole time-of-flight (TOF) mass spectrometer equipped with a traveling wave (T-wave) IMS system (Synapt G2-S, Waters Co., Milford, MA, USA) was used to collect mass spectra. The commercial ion source was removed and the capillary assembly was installed for electrospray ionization in the ambient environment.
4. The parameter settings for the ion mobility mass spectrometer were optimized to achieve maximum separation at the highest ion transmission. All instrument settings were saved in a parameter file. The major parameters included mass range: m/z 50–995, acquisition mode: mobility-TOF, the sensitivity and polarity of ion modes, scan rate: 0.5 s/scan, IMS wave velocity: 650 m/s, and wave height: 40 V.

5. Nitrogen drift gas was supplied at a flow rate of 90 mL/min and a pressure of 3.25 mbar. When the mass spectrometer was operated in MS/MS mode, argon was supplied as a collision gas for performing CID.

2.6 Data Analysis

1. DriftScope 2.8 (Waters Co., Milford, MA, USA) software was used for automatic peak detection, the conversion of drift time (DT) to CCS, and CCS calibration.

3 Methods

To perform capillary microsampling ESI-IMS-MS, customized capillaries are pulled based on the characteristics of cell types and are used for sampling of individual adherent cells. This technique can also be applied for the analysis of suspended cells held by holding pipettes. The glass capillary held by a capillary holder is used to sample individual cells under the observation by an inverted microscope. The capillary is backfilled with electrospray solution and placed in a microelectrode holder. Applying high voltage to the electrical connection on the microelectrode holder produced an electrospray from the cell contents. The generated ions are separated in a T-wave IMS system according to their CCSs and then analyzed by a mass spectrometer based on their mass-to-charge ratios (m/z). Tandem MS is also performed for structural identification of the ions. Time aligned parallel (TAP) fragmentation, where CID is applied after ion separation by IMS, is performed to differentiate and identify isobaric ions with close to identical m/z. The resulting ionic m/z and CCS information are used for metabolite and lipid assignments.

3.1 Preparation of Capillaries for Cell Sampling

1. Initially the program settings of the micropipette puller were adjusted to produce capillaries that achieve optimal cell extraction and reproducible MS signal. A two-step pulling program was set as follows, Step 1: Heat = 574, Pull = 95, Velocity = 40 and Delay = 170, and Step 2: Heat = 564, Pull = 90, Velocity = 70 and Delay = 120 at Pressure = 500. The parameter settings should be varied for other capillary types and filament heating conditions. The optimized parameters were saved to repeatedly produce sampling capillaries on the micropipette puller.
2. Before an experiment, ~20 capillaries were pulled and stored in the pipette storage box to prevent breakage and contamination of the tips.

3.2 CCS Calibration

1. To perform CCS calibration, 1.0 μL poly-DL-alanine solution was backfilled into a capillary through a microloader tip. The capillary was tapped on the side to eliminate air bubbles.

2. The capillary was mounted in the microelectrode holder. This assembly was placed in front of the mass spectrometer inlet orifice with the capillary tip at a distance of ~5 mm.
3. A negative voltage of −1500 V was applied on the microelectrode by a high voltage power supply.
4. The acquisition parameters saved earlier were selected for the mass spectrometer.
5. The acquisition of IMS-MS data for poly-DL-alanine solution was initiated.
6. A CCS calibration curve fitted as a power function was generated using the DriftScope 2.8 software based on the poly-DL-alanine data. The reference CCS values for poly-DL-alanine were obtained from previous publications [8, 11]. The curve with an $R^2 \geq 0.99$ was accepted for the determination of CCS values for unknown ions.

3.3 Capillary Microsampling

1. 0.5× PBS buffer was prepared by mixing equal volume of 1× PBS solution and deionized water. The buffer was warmed to 37 °C in a beads bath for 15 min.
2. Before cell sampling, a 35 mm dish of adherent cells was obtained from the incubator and washed three times by 0.5 mL 0.5× PBS buffer. Then the dish containing 0.5 mL 0.5× PBS was placed on the center of the microscope sample stage.
3. A pulled capillary was inserted in the capillary holder, and they were attached to the micromanipulator at an angle of 45° relative to the microscope stage. The position of the capillary was adjusted to the center of the field of view and at ~4 mm above the cells.
4. The monolayer of cells was brought to the focal plane and a cell of interest was selected by moving the microscope sample stage. Figure 1a shows a microscope image of a HepG2/C3A cell before capillary sampling.
5. The capillary tip controlled by the micromanipulator was carefully moved over the targeted cell and lowered to approach it. In the meantime, the microscope focal plane was adjusted back and forth between the tip and the cell to monitor their relative distance and position.
6. When the tip approached the cell layer, the step size of the micromanipulator was switched from "coarse" to "fine" mode.
7. The capillary was further lowered to slightly touch the cell membrane, and at that point the height of the tip was set as *z*-axis limit for the micromanipulator to prevent tip damage (*see* Fig. 1b and **Note 4**). Then a negative pressure was applied

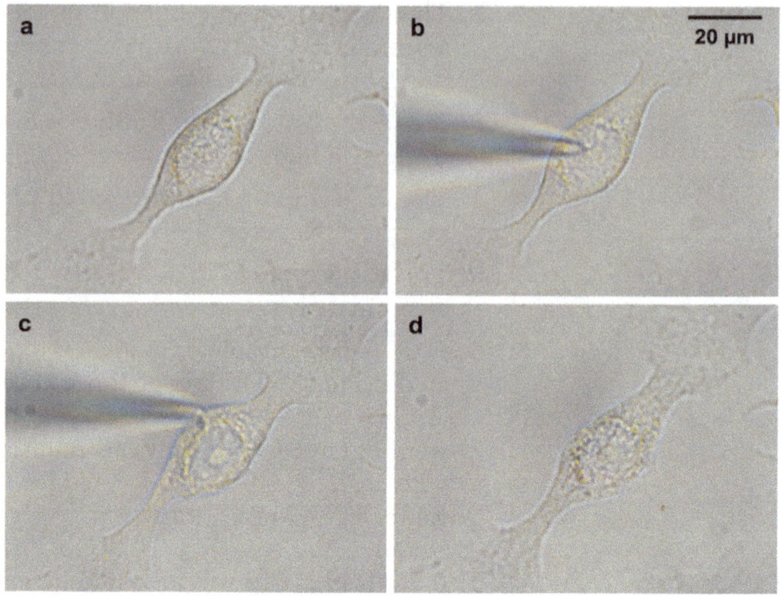

Fig. 1 **(a)** Cell of interest is brought in focus under the microscope. **(b)** Capillary tip is slightly touching the cell membrane before sampling. **(c)** Negative pressure is applied during cell sampling. **(d)** Capillary tip is removed from the cell after sampling

using a syringe attached to the back of the capillary holder to extract the cell contents (*see* Fig. 1c).

8. After sampling the cell contents, the tip was removed from the dish (*see* Fig. 1d).
9. The cells were returned to the incubator.

3.4 Single Cell ESI-IMS-MS

1. The capillary was removed from the capillary holder, and back-filled with 1.0 μL electrospray solution. To remove air bubbles, the capillary was gently flicked on the side.
2. The capillary was placed in the microelectrode holder and the platinum wire came in contact with the electrospray solution. The capillary tip was aligned with the orifice of the mass spectrometer at a distance of ~5 mm.
3. The data acquisition by the mass spectrometer was initiated using the previously saved parameters.
4. To generate ions from the sampled cell contents, a high voltage of −1500 V was applied to the microelectrode. Typically, during the first 5–10 s of the electrospray, cell related signal was collected, whereas the electrospray background was acquired for the following 10 s (*see* **Note 5**).

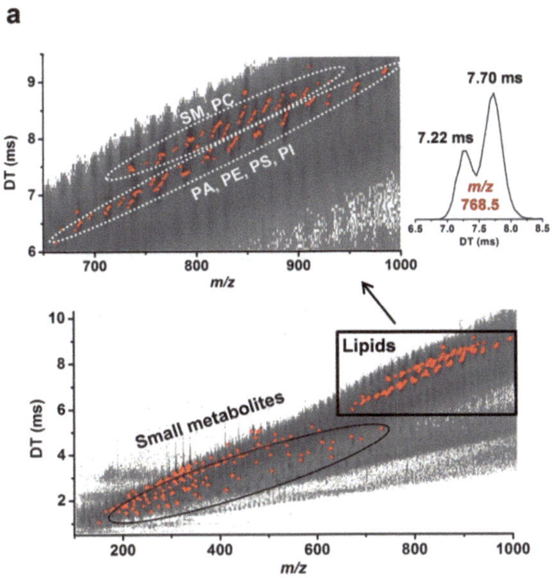

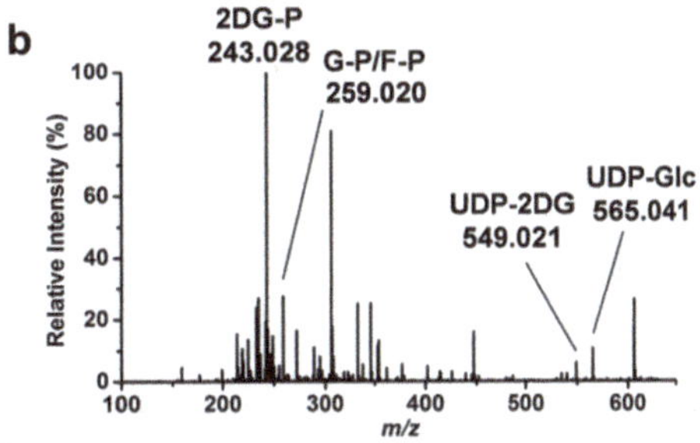

c

Identification	m_{meas} (mDa)	Δm (mDa)	CCS_{meas} (Å^2)	ΔCCS (Å^2)
2-DG-P	243.028	0.5	142.5	
G-P/F-P	259.020	-2	142.6	-0.4
UDP-2DG	549.046	-7	203.2	
UDP-Glc	565.041	-7	207.2	3.2
PE(18:0/20:3)	768.550	-5	284.0	
PC(16:0/16:0)	768.520	-11	295.2	

Fig. 2 (a) At the bottom, DT vs. *m/z* plot for a cell treated with 5 mM 2-DG for 1 h with the detected ions marked by red dots. Metabolite and lipid species are separated and highlighted in distinct regions. On the top, the zoomed DT vs. *m/z* plot in the lipid region shows sphingomyelin (SM) and phosphatidylcholine (PC) lipids separate from phosphatidic acid (PA), phosphatidylethanolamine (PE), phosphatidylserine (PS), and phosphatidylinositol (PI) lipid species spanning the same *m/z* range. A DT distribution of a pair of separated isobaric species with nominal *m/z* 768.5 is shown on the right. **(b)** Mass spectrum of the 2-DG-treated cell shows two new biotransformation products, 2-DG phosphate (2-DG-P) and UDP-2-deoxyglucose (UDP-2DG). **(c)** Table lists selected metabolites identified based on accurate mass measurement, CCS values, and tandem mass spectrometry (G-P/F-P = glucose phosphate or fructose phosphate, UDP-Glc = UDP-glucose). The Δm values indicate the deviation from calculated accurate mass, whereas ΔCCS stands for the difference from literature values

3.5 Data Processing and Metabolite Identification

1. A DT vs. *m/z* plot representing a three-dimensional dataset comprised of ion abundances, DT, and *m/z* information was collected for the analysis of each single cell. DriftScope 2.8 software was used to visualize and process the data. Figure 2a shows a DT vs. *m/z* plot of a cell treated by 2-DG at 5 mM for 1 h. The top panel shows that lipid classes are partially separated within an *m/z* range. An example of the DT distribution for isobaric ions with nominal *m/z* 768.5 is shown on the right.
2. Ions of interest in different regions on the DT vs. *m/z* plot were selected and exported to the MassLynx 4.1 software (Waters Co., Milford, MA) to generate and process the corresponding mass spectra. Figure 2b shows the exported mass spectrum corresponding to small metabolites. Two biotransformation products produced in the cell in response to 2-DG treatment, 2-DG phosphate (2-DG-P) and UDP-2-deoxyglucose (UDP-2DG) were detected.

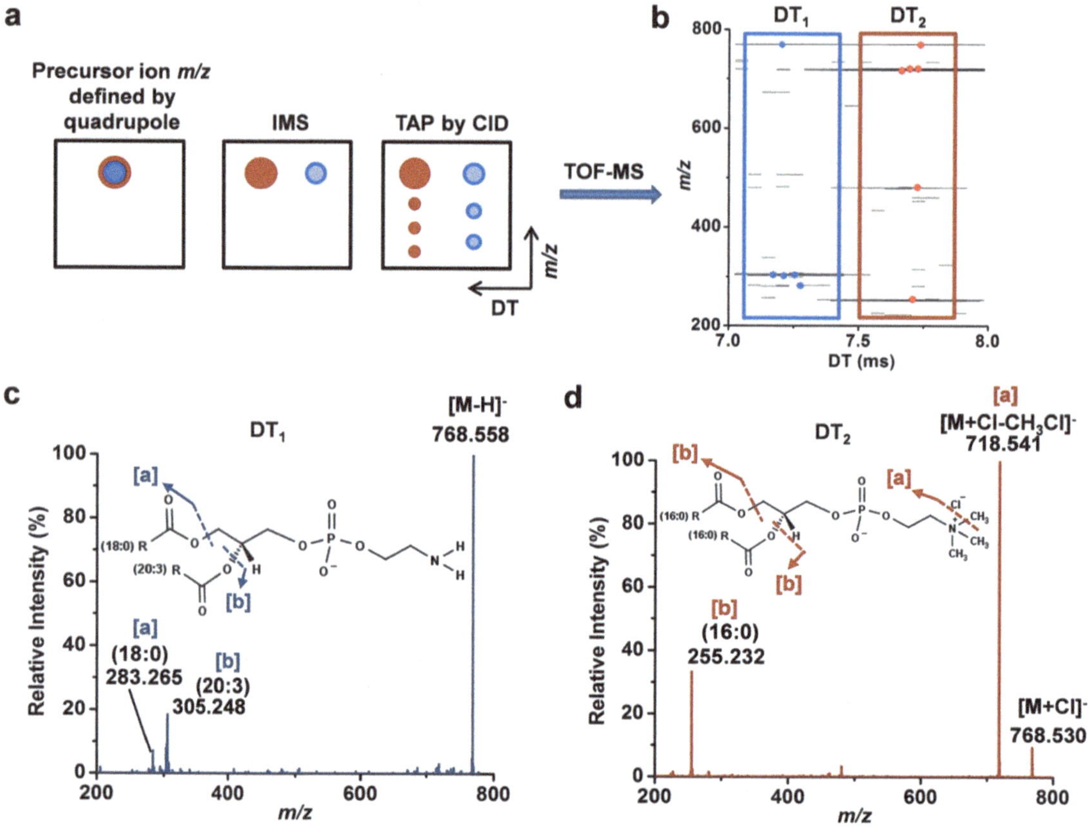

Fig. 3 **(a)** Schematic representation of the TAP fragmentation procedure for distinction and identification of isobaric ions. Precursor ion *m/z* is selected by quadrupole mass analyzer. The isobaric ions with defined *m/z* are resolved by IMS, followed by CID to produce drift time correlated fragment ions. **(b)** Single cell DT vs. *m/z* plot for TAP fragmentation of isobaric ions with nominal *m/z* 768.5. The precursor ion at $7.0 < DT_1 < 7.4$ ms and its corresponding fragment ions are marked by blue dots and framed in a blue rectangle, whereas the ion at $7.5 < DT_2 < 7.9$ ms and its fragment ions are marked by red dots and framed in a red rectangle. Tandem mass spectra for **(c)** the DT_1 ion and **(c)** the DT_2 ion with the fragmentation patterns are shown in the insets. DT_1 and DT_2 ions are identified as [PE (18:0/20:3)-H]$^-$ and [PC (16:0/16:0) + Cl]$^-$, respectively

3. The determination of CCS values for the unknown ions was based on the CCS calibration curve for poly-DL-alanine produced by the DriftScope 2.8 software. The measured *m/z* and CCS values were used for tentative assignment of the unknown metabolites by searching metabolomics databases, e.g., Human Metabolome Database (http://www.hmdb.ca/), and previously published CCS values, respectively [11, 12] (*see* Fig. 2c).
4. For further structural identification of the unknown ions, TAP fragmentation was performed and tandem mass spectra were acquired in mobility-TOF MS/MS mode. Before IMS, the nominal *m/z* of the unknown ion was set as the precursor mass. Following IMS, the ions were subjected to CID with the collision energy scanned between 10 and 40 eV. Figure 3a

shows a representation of TAP fragmentation to distinguish and identify two isobaric ions. Figure 3b shows an example of a DT vs. *m/z* plot for TAP fragmentation of a pair of isobaric ions with nominal *m/z* 768.5. Distinct tandem mass spectra for individual isobaric ions were derived (*see* Fig. 3c and d). According to the fragmentation patterns, they were assigned as [PE (18:0/20:3)-H]$^-$ and [PC(16:0/16:0) + Cl]$^-$, respectively.

4 Notes

1. Depending on the cell line and growth rate, an appropriate density of cells should be seeded on the Petri dish to make sure that individual cells could be easily distinguished for sampling.
2. Direct contact with the sharp end of the capillary tip can cause injury and should be avoided. The capillaries should be disposed in sharp containers.
3. Direct contact with the high voltage connections can cause electric shock or death. All the electrical components should be insulated and shielded. Do not touch the electrical components when the high voltage power supply is turned on.
4. When touching the dish bottom, the capillary tip might break. On such occasions, a large amount of PBS solution enters the capillary. To avoid PBS interference in the mass spectra, a new tip should be used.
5. While approaching the cell, the PBS solution may enter the capillary tip and introduce interference during analysis. Mass spectra of the PBS solution are separately acquired for background subtraction.

Acknowledgements

The authors acknowledge the financial support from the U.S. National Science Foundation under Grant No. CHE-1152302.

References

1. Walker BN, Antonakos C, Retterer ST, Vertes A (2013) Metabolic differences in microbial cell populations revealed by Nanophotonic ionization. Angew Chem Int Ed 52:3650–3653
2. Ibanez AJ, Fagerer SR, Schmidt AM, Urban PL, Jefimovs K, Geiger P, Dechant R, Heinemann M, Zenobi R (2013) Mass spectrometry-based metabolomics of single yeast cells. Proc Natl Acad Sci U S A 110:8790–8794

3. Zenobi R (2013) Single-cell metabolomics: analytical and biological perspectives. Science 342:1243259
4. Shrestha B, Vertes A (2009) In situ metabolic profiling of single cells by laser ablation electrospray ionization mass spectrometry. Anal Chem 81:8265–8271
5. Svatos A (2011) Single-cell metabolomics comes of age: new developments in mass spectrometry profiling and imaging. Anal Chem 83:5037–5044
6. Wilson ID, Plumb R, Granger J, Major H, Williams R, Lenz EA (2005) HPLC-MS-based methods for the study of metabonomics. J Chromatogr B 817:67–76
7. Pitt JJ (2009) Principles and applications of liquid chromatography-mass spectrometry in clinical biochemistry. The Clinical biochemist Reviews / Australian Association of Clinical Biochemists 30:19–34
8. Bush MF, Campuzano IDG, Robinson CV (2012) Ion mobility mass spectrometry of peptide ions: effects of drift gas and calibration strategies. Anal Chem 84:7124–7130
9. Lanucara F, Holman SW, Gray CJ, Eyers CE (2014) The power of ion mobility-mass spectrometry for structural characterization and the study of conformational dynamics. Nat Chem 6:281–294
10. Shrestha B, Vertes A (2014) High-throughput cell and tissue analysis with enhanced molecular coverage by laser ablation electrospray ionization mass spectrometry using ion mobility separation. Anal Chem 86:4308–4315
11. Paglia G, Williams JP, Menikarachchi L, Thompson JW, Tyldesley-Worster R, Halldorsson S, Rolfsson O, Moseley A, Grant D, Langridge J, Palsson BO, Astarita G (2014) Ion mobility derived collision cross sections to support metabolomics applications. Anal Chem 86:3985–3993
12. Paglia G, Angel P, Williams JP, Richardson K, Olivos HJ, Thompson JW, Menikarachchi L, Lai S, Walsh C, Moseley A, Plumb RS, Grant DF, Palsson BO, Langridge J, Geromanos S, Astarite G (2015) Ion mobility-derived collision cross section as an additional measure for lipid fingerprinting and identification. Anal Chem 87:1137–1144
13. Fenn LS, McLean JA (2011) Structural resolution of carbohydrate positional and structural isomers based on gas-phase ion mobility-mass spectrometry. PCCP 13:2196–2205
14. Zhang L, Foreman DP, Grant PA, Shrestha B, Moody SA, Villiers F, Kwak JM, Vertes A (2014) In situ metabolic analysis of single plant cells by capillary microsampling and electrospray ionization mass spectrometry with ion mobility separation. Analyst 139:5079–5085
15. Zhang L, Sevinsky CJ, Davis BM, Vertes A (2018) Single-cell mass spectrometry of subpopulations selected by fluorescence microscopy. Anal Chem 90:4626–4634
16. Zhang L, Khattar N, Kemenes I, Kemenes G, Zrinyi Z, Pirger Z, Vertes A (2018) Subcellular peptide localization in single identified neurons by capillary microsampling mass spectrometry. Sci Rep 8, 12227
17. Zhang L, Vertes A (2018) Single-cell mass spectrometry approaches to explore cellular heterogeneity. Angew Chem Int Ed 57:4466–4477

Chapter 3

Analysis of Lipids in Single Cells and Organelles Using Nanomanipulation-Coupled Mass Spectrometry

Mandy S. Phelps and Guido F. Verbeck

Abstract

The ability to discriminately analyze the chemical constituents of single cells and organelles is highly sought after and necessary to establish true biomarkers. Some major challenges of individual cell analysis include requirement and expenditure of a large sample of cells as well as extensive extraction and separation techniques. Here, we describe methods to perform individual cell and organelle extractions of both tissues and cells in vitro using nanomanipulation coupled to mass spectrometry. Lipid profiles display heterogeneity from extracted adipocytes and lipid droplets, demonstrating the necessity for single cell analysis. The application of these techniques can be applied to other cell and organelle types for selective and thorough monitoring of disease progression and biomarker discovery.

Key words Nanomanipulation, Mass spectrometry, Single cell, Nanoelectrospray, MALDI, Lipidomics, DOMS

1 Introduction

Lipids are vital to cell physiology, serving functions in cellular membranes, energy storage, and cellular signaling. Their roles and pathways in metabolism have warranted several implications relating to metabolic disorders and an entire field dedicated to their studies, lipidomics [1, 2]. Small sample volumes and analytical limitations for lipidomics experiments have resulted in several extraction and separation protocols, which typically require expenditure of a large group of cells from both tissues and cell cultures. While useful, the results procured represent an inherent average of cell population and loss of spatial chemical information. The need for individual cell and organelle analyses has been made obvious as improvements to analytical tools have revealed cellular and subcellular heterogeneity [3–9]. Many single cell analytical techniques have been introduced to negate the problem of population cell

Bindesh Shrestha (ed.), *Single Cell Metabolism: Methods and Protocols*, Methods in Molecular Biology, vol. 2064,
https://doi.org/10.1007/978-1-4939-9831-9_3,

averaging [10–21], but have significant drawbacks including consumption of entire cell sample [14–17], lack of chemical structural information [18] or targeted specificity [19, 21], and lengthy purification, separation, and digestion requirements [14, 16, 19, 20].

We describe alternative solutions to these issues in protocols for obtaining both individual cell and organelle lipid profiles by using a nanomanipulation platform coupled to mass spectrometry utilizing multiple ionization sources. The nanomanipulator our group has developed is ideal for discriminately extracting and manipulating cellular components [22–25] and small particles [26–29], due to having precise translational resolution on the nanometer scale. The technique avoids separation and purification steps, and is useful for both tissues and cells in culture, with adipocytes being the subject for the following methods. The specificity and non-invasiveness of the nanomanipulator enables a rapid way to obtain and accurately analyze targeted biomaterials of interest, while leaving the rest of the tissue section or cell culture intact for further experiments. Described here are the protocols successfully used to perform single-cell tissue extractions, in vitro whole cell digestions, and in vitro organelle extractions stemming from refs [23–25] respectively.

2 Materials

2.1 Tissue Extractions

1. CM 1850 Cryomicrotome for tissue sectioning (Leica Microsystems, Buffalo Grove, IL).
2. Glass slides (cat. no. 12-550-15, Fisher Scientific, Fair Lawn, NJ).
3. P-2000 CO_2-laser tip puller system (Sutter Instruments, Novato, CA).
4. Solid quartz rods (cat. no. QR100-10, Sutter Instruments, Novato, CA).
5. Econo12 PicoTip Nanospray Emitters (New Objective, Woburn, MA).
6. L200 nanomanipulator (DCG Systems, Inc., Fremont, CA).
7. Methanol (MeOH) (Optima LC/MS Fisher Scientific, Fair Lawn, NJ).
8. Chloroform ($CHCl_3$) (CHROMASOLV® for HPLC Sigma Aldrich, St. Louis, MO).
9. Ammonium acetate (NH_4OAc) (cat. no. A1542, Sigma Aldrich, St. Louis, MO).

10. Pressure injector (MicroData Instrument Inc., S. Plainfield, NJ) using nitrogen gas.
11. Breast cancer and adjacent healthy tissue (Cureline, Inc., South San Francisco, CA).
12. AZ100 microscope (Nikon, Melville, NJ).

2.2 Cell Culture Extractions

1. Primary human skin fibroblasts (Coriell Institute, Camden, NJ).
2. 3T3-L1 murine fibroblasts (ATCC, Manassas, VA).
3. 35 mm round glass bottom culture dishes (MatTek, Ashland, MA).
4. Penicillin/streptomycin (cat. no. SV30010, Thermo Scientific, San Jose, CA).
5. Humalog insulin (Lilly, Indianapolis, IN).
6. TE2000 inverted microscope (Nikon, Melville, NJ).
7. Dulbecco's modified eagle's medium (DMEM) (cat. no. D6429), Fetal bovine serum (FBS) (cat. no. F6178), 3-isobutyl-1-methylxanthine (IBMX) (cat. no. I5879), Dexamethasone (cat. no. D4902), Indomethacin (cat. no. I7378), 2,5-Dihydroxybenzoic acid (DHB) (cat. no. 149357), Palmitic acid (cat. no. P5585), Oleic acid (cat. no. O1383), Palmitic acid-1-^{13}C (cat. no. 292125), and Oleic acid-1-^{13}C (cat. no. 490423) (all from Sigma-Aldrich, St. Louis, MO).
8. Phosphate buffered saline (PBS) (Fisher Scientific, Fair Lawn, NJ).
9. Acetonitrile (ACN) (Optima LC/MS Fisher Scientific, Fair Lawn, NJ).
10. Ultrapure 18.2 MΩ water (Milli-Q®, Billerica, MA) (H_2O).

2.3 Mass Spectrometry

1. NSI ionization source (Proxeon Biosystems, Odense, Denmark).
2. LCQ DECA XP Plus mass spectrometer (Thermo Scientific, San Jose, CA).
3. LTQ XL mass spectrometer (Thermo Scientific, San Jose, CA).
4. MNL 100337 nm N_2 laser (Lasertechnik, Berlin, Germany).
5. MALDI-LTQ-XL-Orbitrap (Thermo Scientific, San Jose, CA).

2.4 MS Data Processing

1. Xcalibur v 2.2 (Thermo Scientific, San Jose, CA).
2. ImageQuest v 1.01 (Thermo Scientific, San Jose, CA).
3. PSI-Plot (Poly Software International, Pearl River, NY).

3 Methods

3.1 Single Cell Tissue Extraction

1. Section tissue in 80 μm thick (*see* **Note 1**) sections using a cryomicrotome and place onto glass slides directly before extraction experiments.
2. Prepare extraction solvent consisting of 2:1 $CHCl_3$:MeOH (v/v) with 0.1% NH_4OAc. This solvent should be prepared and stored using glass on the day of extraction (*see* **Note 2**). Backfill 10 μL of solvent into the nanospray emitter and place into a nanopositioner connected to the pressure injector for aspiration and injection of solvent.
3. A quartz rod can be pulled using a P-2000 tip puller system to create an ~8 μm probe using a pull setting of 55, heat setting of 700, delay setting of 130, and velocity setting of 55. This is used to create a puncture in the targeted adipocyte before nanoextraction to avoid both clogging of the nanospray emitter and unwanted membrane chemistry. Quartz probes for puncturing of tissue can be made prior to extraction if kept in a sealed container. Place the pulled quartz probe into a second nanopositioner (Fig. 1a) and glass slide with tissue sections onto the microscope.
4. After locating the adipocyte, direct the nanopositioner holding the quartz probe using the joystick to the membrane and within using the *z* direction. Upon removal of the quartz

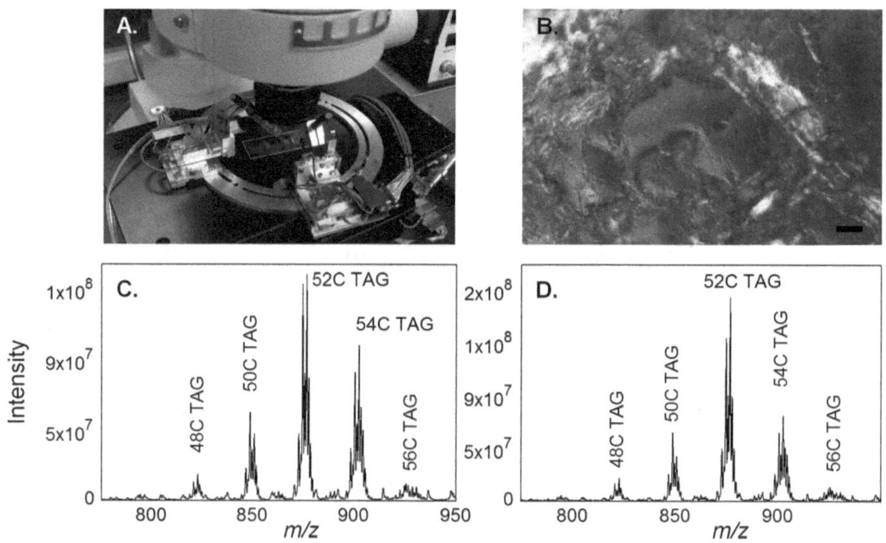

Fig. 1 (**a**) Image showing nanomanipulator station mounted onto the AZ100 microscope equipped with two nanopositioners for use in puncturing and tissue extraction. (**b**) Adipocyte in healthy human breast tissue post quartz rod puncture, 25 μm scale bar. (**c**) TAG profile obtained from healthy breast tissue adipocyte extraction. (**d**) TAG profile obtained from adjacent cancerous adipocyte extraction

probe from the adipocyte, a distinct puncture site should be observed (Fig. 1b).

5. Following confirmation of puncture site, transition the adjacent nanopositioner holding the nanospray emitter and extraction solvent directly into the puncture created by the quartz probe. For tissue extractions, use the injection control for 0.5 s at a pressure of 5 psi to release extraction solvent into the adipocyte and disrupt the tip-to-tissue interface. Following injection, aspirate the injected solvent and digested lipid content obtained using a pressure of 20 psi from 1 to 5 s.
6. Remove the nanospray emitter from the cell using the *z* direction of the nanopositioner. Allow at least 5 min for extracted content and solvent to mix before transferring emitter to the nanospray source for mass analysis (*see* **Note 3**).
7. Follow nanoelectrospray ionization protocol, Subheading 3.5. Figure 1 shows an example of the results obtained from extraction of a healthy adipocyte (Fig. 1c) in comparison to a cancerous adipocyte (Fig. 1d) from human breast tissue. *See* ref [23] for more information.

3.2 Adipogenic Differentiation

This differentiation protocol can be applied to both 3T3-L1 murine and primary human fibroblast lines adapting established methods [30–32] and incorporating ^{13}C labeled and non-labeled fatty acids in the 3T3-L1 cell line for uptake monitoring.

1. Seed fibroblasts into 35 mm round glass bottom culture dishes at 15,000 cells per dish in 3 mL of medium. Growth medium consists of DMEM containing 10% FBS and 1% penicillin/streptomycin, and should be replaced every 2–3 days. Store cultures in a 5% CO_2 incubator at 37 ° C.
2. When cells reach 100% confluency (~2 weeks), remove growth medium, rinse with 1× PBS, and replace with 3 mL of differentiation medium. Prepare differentiation medium by incorporating 10 μg/mL insulin, 1 μM dexamethasone, 0.2 mM indomethacin, and 0.5 mM IBMX into growth medium, and also replace every 2–3 days. Additionally, 100 μM of ^{13}C labeled or non-labeled oleic and palmitic acid can be added to differentiation medium to encourage lipid droplet growth [31] and perform tracking studies of specific fatty acid uptake.
3. Lipid droplets of varying sizes should be visualized within 10–14 days of addition of differentiation medium.

3.3 Whole Cell Digestion in Culture

1. Prepare culture dish for nanomanipulation and extraction by removing growth medium from culture dishes, rinse with 1 mL of 1× PBS, and add 1 mL of 1× PBS.

2. Backfill 10 μL of extraction solvent into a nanospray emitter as directed in Subheading 3.1, **step 2** (*see* **Note 4**).
3. Place culture dish containing adipocytes onto inverted microscope, focusing on targeted cell for digestion and transitioning the nanopositioner holding the emitter directly above and just in contact with (*see* **Note 5**) the cell using the joystick.
4. Use the inject function of the pressure injector for 0.1–0.5 s at 3 psi to create a solvent bubble to cover the whole adipocyte. Dissolvation of cell and lipid droplets within should be visually confirmed before aspiration back into the emitter occurs (*see* **Note 6**).
5. Perform aspiration of solvent and dissolved adipocyte contents using 18 psi pressure until visual confirmation of retracting solvent bubble is complete (*see* **Note 7**).
6. Remove the nanospray emitter from the culture dish using the *z* direction of the nanopositioner, and allow to mix for 5 min.
7. Follow nanoelectrospray ionization protocol, Subheading 3.5. An example of the results obtained is displayed in Fig. 2 *see* ref [24] for detailed information.

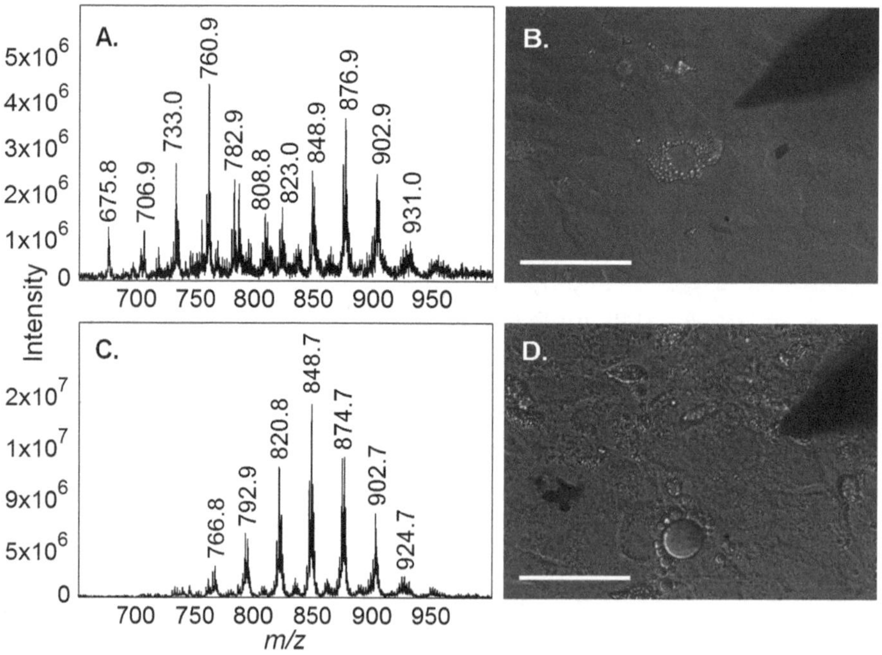

Fig. 2 (**a**) Lipid profile obtained from performing whole cell digestion of a differentiated human adipocyte containing small lipid droplets shown in (**b**) Corresponding lipid profile (**c**) obtained from digesting an adipocyte containing large lipid droplets from the same culture dish shown in (**d**). Scale bar 25 μm

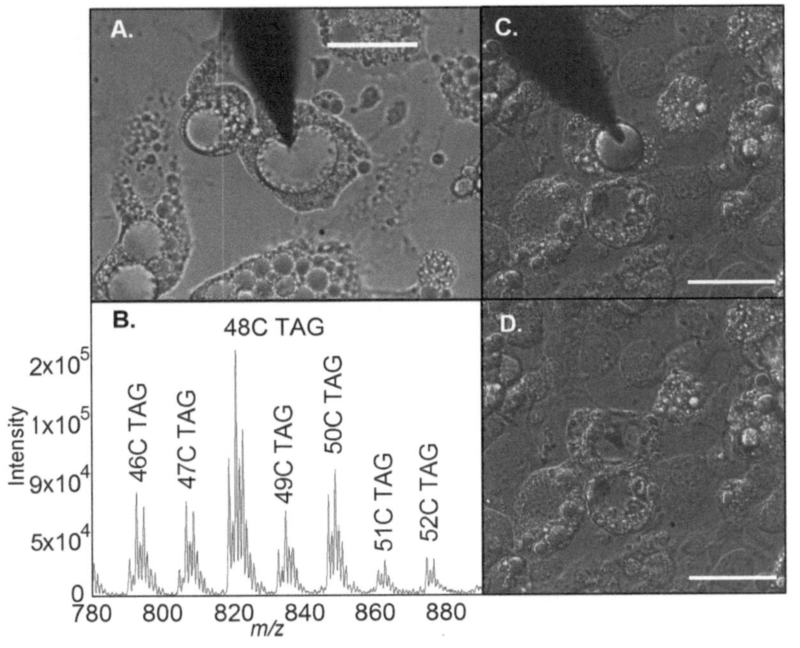

Fig. 3 (**a**) Nanospray emitter within individual lipid droplet of 3T3-L1 adipocyte, performing aspiration of organelle content. (**b**) NSI-MS TAG profile obtained from extraction of the lipid droplet. (**c**) Nanospray emitter in contact with one lipid droplet before extraction. (**d**) Post extraction image of organelle shown in (**c**) 50 μm scale bars

3.4 Direct Organelle Extractions

1. Prepare adipocytes for extraction and nanospray emitter as directed in Subheading 3.3, **step 1** and Subheading 3.1, **step 2**.
2. Direct nanospray emitter using joystick-controlled nanopositioner to targeted adipocyte containing lipid droplets of interest. Using the *z* direction of the joystick, position the tip of the emitter directly into contact with the center of the targeted lipid droplet.
3. Gradually and carefully continue increasing the depth of the *z* direction until the tip of the emitter breaks the surface tension of the lipid droplet membrane (Fig. 3a) and is visually observed within the targeted lipid droplet (*see* **Note 8**).
4. Use the aspiration function of the pressure injector with 20 psi pressure to retrieve organelle content into the nanospray emitter, stopping when the lipid droplet is no longer observed in the microscope (Fig. 3c→d).
5. Remove the nanospray emitter with extracted organelle content out of the culture dish and allow to mix for 5 min.
6. Extracted cell content can either be analyzed using nanoelectrospray ionization (Subheading 3.5, example of results shown

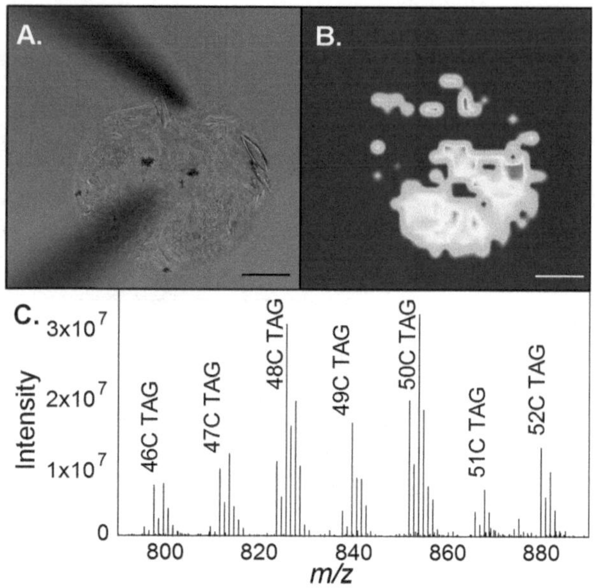

Fig. 4 (**a**) Co-deposition spot consisting of extracted lipid droplet content followed by DHB matrix deposition, with both nanospray emitters shown above. (**b**) MSI image created from the total ion count of the co-deposited spot. 100 μm scale bars. (**c**) TAG profile obtained from lipid droplet extraction with 100,000 resolution using MALDI

in Fig. 3b) or by MALDI using a co-deposition method (Subheading 3.6, example of results shown in Fig. 4c).

3.5 Nanoelectrospray-Direct Organelle Mass Spectrometry (NSI-DOMS)

1. Following extraction procedures, transfer nanospray emitter to NSI source, positioning the tip of the emitter slightly offset from the inlet of the heated capillary (*see* **Note 9**).
2. Set instrumental conditions as follows:
   (a) LCQ DECA XP Plus—Source voltage 2.5 kV, capillary temperature 250 °C, capillary voltage 3.5 V, positive mode, mass range of *m/z* 500–1000 (*see* **Note 10**).
   (b) LTQ XL—Source voltage 2.5 kV, capillary temperature 250 °C, capillary voltage 9 V, positive mode, mass range of *m/z* 500–1000.

3.6 Matrix-Assisted Laser/Desorption Ionization (MALDI-DOMS)

1. Prepare fresh matrix solution of 20 mg/mL DHB in 3:2 ACN: H_2O (v/v), and backfill 10 μL into a second nanospray emitter for deposition of matrix.
2. Following extraction of organelle content (Subheading 3.4), remove culture dish from inverted microscope and replace with a clean glass slide.
3. Direct nanopositioner containing extracted organelle content to come just into contact with the glass slide (*see* **Note 11**).

4. Pulse-spot extracted content at 1.5 psi using the pressure injector for 0.3 s.
5. Allow spotted content to dry, and repeat ~20 times to create a concentrated spot of analyte on the glass slide. Transition the nanopositioner away from the area.
6. Direct the second nanopositioner containing matrix solution directly above and in contact with the analyte spot created.
7. Use the inject function of the pressure injector to spot matrix on top of the extracted lipid content at 1.5 psi until the original spot is covered with matrix. Allow to dry and observe matrix crystals formed (Fig. 4a) (*see* **Note 12**).
8. Place glass slide into the glass slide adapter for MALDI mass analysis.
9. Inject the MALDI plate into the instrument, and locate the co-deposited spot by scanning the glass slide with the camera.
10. Set instrumental conditions as follows: 50 μm laser raster motion step size, 10 laser shots per spectra, mass range of *m/z* 700–1000, 12 μJ energy per laser shot, acquired in positive ion mode. Set resolution of instrument to 100,000 (at *m/z* 400). By acquiring a scan of the co-deposited spot, a mass spectrometry image (MSI) (Fig. 4b) is created which allows for localization of selected lipids. An example of the mass spectrum obtained is shown in Fig. 4c. *See* ref [25] for more information.

4 Notes

1. For MALDI imaging of tissues, a much thinner tissue section is typically desired. However, for nanomanipulation purposes, a thicker section is needed to ensure that drying out of the section does not occur. A thickness of 80 μm allowed for several hours of extractions before the sample dried out compared to 20, 30 and 40 μm sections. Additionally, the added thickness of tissue provides a more pliable surface for the emitter and probe to come into contact with before reaching the glass surface.
2. This extraction solvent is suitable for multiple classes of lipids, such as diacylglycerols, triacylglycerols, and some glycerophosphocholine species. The solvent can and should be adjusted for targeting of different cellular components.
3. If the extraction is taken immediately to the ionization source without sufficient time for mixing, the time for mass analysis is greatly decreased (seconds vs. minutes). By allowing the extracted analyte and solvent time to mix, more mass analysis time is procured for the sample. This provides a more

representative lipid profile and enables MS/MS experiments to be performed. However, the limited volume of extracted cell content is certainly not completely resolved with additional time for mixing. An average of 2 min analysis time was obtained with individual cell extractions from human breast tissue.

4. When backfilling an emitter, it is important to avoid air bubbles. If a bubble is present at the tip opening, it becomes difficult to create a tip-to-surface interface for successful injection and aspiration of analyte. Additionally, if an air bubble is present in the emitter during NSI, detection of cell and organelle content may not occur or a drop in signal will appear in the chromatogram as the air bubble reaches the emitter tip opening.
5. The fibroblasts and adipocytes described in this work form a very thin, adhered layer to the glass of the culture dish. If the emitter is directed too quickly or deeper than just into contact with the cell, the tip will come into contact with the glass and possibly break or affect the tip opening diameter. Decreasing the speed of the *z* direction on the nanomanipulator can help with this problem, as well as changing the material of the emitter used (quartz vs. glass).
6. The tip opening diameter of emitters can vary, especially of those created in-house or those that may have come into contact with the glass of the cell culture dish. It is important to adjust the pressure parameter on the injector to accommodate diameter size. Smaller tip opening diameters often require higher injection pressures, while larger tip openings require lower injection pressures to be used effectively. If the pressure is not changed for a larger tip opening, the injection of solvent will create a much larger solvent bubble than desired, reaching multiple cells and negating the purpose of performing individual cell extractions.
7. Due to the immiscibility of the extraction solvent with PBS, the solvent bubble created from injection will remain attached to the nanospray emitter during the dissolvation event. During aspiration, the solvent bubble will retract back into the emitter, becoming smaller and smaller until the entirety of the injected solvent is collected back into the emitter. It is at this point when the user can stop the aspiration event. The time using this function will vary based on the size of the solvent bubble created.
8. The positioning of the tip opening at the center of the lipid droplet is crucial for successful entering of the emitter. If positioned on an outer edge of the droplet, the emitter will not break the surface tension and enter for extraction, it will merely push the organelle in motion with the tip. Just as the emitter

enters the lipid droplet, stop increasing depth using the z direction to avoid contact with the glass below.

9. If the emitter is positioned directly center toward the opening of the capillary of the mass spectrometer, a concentrated amount of solvent molecules with varying size distributions will enter the mass spectrometer [33], affecting sensitivity, buildup of residue on the capillary, and signal intensity [34].
10. This mass range is suitable for encompassing multiple lipid classes in positive ion mode. The mass range can be adjusted to observe targeted lipid classes, such as triacylglycerols (m/z 750–950), or when utilizing tandem mass spectrometry to view product ions.
11. Owing to capillary action, extracted content and solvent may release from the emitter when coming into contact with the glass slide. If this happens, direct the emitter away from the slide using the z direction control, and allow spotted content to dry. Continue bringing the emitter into contact with the glass in the same spot to create a concentrated area of analyte.
12. If matrix crystals do not form uniformly over concentrated analyte spot, bring the nanospray emitter containing matrix back into contact with the spot and inject additional matrix over the area. If left for too long, the matrix solution within the nanospray emitter can crystallize toward the tip opening and block/clog the injection function. If this happens, obtain and backfill a new nanospray emitter with fresh matrix solution.

References

1. Teo CC, Chong WPK, Tan E, Basri NB, Low ZJ, Ho YS (2015) Advances in sample preparation and analytical techniques for lipidomics study of clinical samples. Trends Anal Chem 66:1–18
2. Wenk MR (2005) The emerging field of lipidomics. Nat Rev Drug Discov 4:594–610
3. Oikawa A, Saito K (2012) Metabolite analyses of single cells. Plant J 70:30–38
4. Trouillon R, Passarelli MK, Wang J, Kurczy ME, Ewing AG (2013) Chemical analysis of single cells. Anal Chem 85:522–542
5. Borland LM, Kottegoda S, Phillips KS, Allbritton NL (2008) Chemical analysis of single cells. Annu Rev Anal Chem 1:191–227
6. Rubakhin SS, Lanni EJ, Sweedler JV (2013) Progress toward single cell metabolomics. Curr Opin Biotechnol 24:95–104
7. Zenobi R (2013) Single-cell metabolomics: analytical and biological perspectives. Science 342:1201
8. Ide Y, Waki M, Ishizaki I, Nagata Y, Yamazaki F, Hayasaka T, Masaki N, Ikegami K, Kondo T, Shibata K, Ogura H, Sanada N, Setou M (2014) Single cell lipidomics of SKBR-3 breast cancer cells by using time-of-flight secondary-ion mass spectrometry. Surf Interface Anal 46:181–184
9. Speicher MR (2013) Single-cell analysis: toward the clinic. Genome Med 5:74/71–74/73. 73 pp
10. Espina V, Wulfkuhle JD, Calvert VS, Van Meter A, Zhou W, Coukos G, Geho DH, Petricoin EF III, Liotta LA (2006) Laser capture microdissection. Nat Protoc 1:586–603
11. Masujima T (2009) Live single-cell mass spectrometry. Anal Sci 25:953–960
12. Ellis SR, Ferris CJ, Gilmore KJ, Mitchell TW, Blanksby SJ, in het Panhuis M (2012) Direct lipid profiling of single cells from inkjet printed microarrays. Anal Chem 84:9679–9683
13. Lanni EJ, Rubakhin SS, Sweedler JV (2012) Mass spectrometry imaging and profiling of single cells. J Proteome 75:5036–5051

14. Rubakhin SS, Garden RW, Fuller RR, Sweedler JV (2000) Measuring the peptides in individual organelles with mass spectrometry. Nat Biotechnol 18:172–175
15. Chandra S, Lorey DR II (2001) SIMS ion microscopy in cancer research: single cell isotopic imaging for chemical composition, cytotoxicity and cell cycle recognition. Cell Mol Biol 47:503–518
16. Whittal RM, Keller BO, Li L (1998) Nanoliter chemistry combined with mass spectrometry for peptide mapping of proteins from single mammalian cell lysates. Anal Chem 70:5344–5347
17. Hebbar S, Schulz WD, Sauer U, Schwudke D (2014) Laser capture microdissection coupled with on-column extraction LC-MSn enables lipidomics of fluorescently labeled drosophila neurons. Anal Chem 86:5345–5352
18. Wu H, Volponi JV, Oliver AE, Parikh AN, Simmons BA, Singh S (2011) In vivo lipidomics using single-cell raman spectroscopy. Proc Natl Acad Sci U S A 108:3809–3814, S3809/3801-S3809/3811
19. Aerts JT, Louis KR, Crandall SR, Govindaiah G, Cox CL, Sweedler JV (2014) Patch clamp electrophysiology and capillary electrophoresis-mass spectrometry metabolomics for single cell characterization. Anal Chem 86:3203–3208
20. Garden RW, Shippy SA, Li L, Moroz TP, Sweedler JV (1998) Proteolytic processing of the Aplysia egg-laying hormone prohormone. Proc Natl Acad Sci U S A 95:3972–3977
21. Lorenzo TM, Mizuno H, Tsuyama N, Harada T, Masujima T (2012) In situ molecular analysis of plant tissues by live single-cell mass spectrometry. Anal Chem 84:5221–5228
22. Horn PJ, Ledbetter NR, James CN, Hoffman WD, Case CR, Verbeck GF, Chapman KD (2011) Visualization of lipid droplet composition by direct organelle mass spectrometry. J Biol Chem 286:3298–3306
23. Phelps M, Hamilton J, Verbeck GF (2014) Nanomanipulation-coupled nanospray mass spectrometry as an approach for single cell analysis. Rev Sci Instrum 85:124101–124105
24. Phelps MS, Verbeck GF (2015) A lipidomics demonstration of the importance of single cell analysis. Anal Methods 7:3668–3670
25. Phelps MS, Sturtevant D, Chapman KD, Verbeck G (2016) Nanomanipulation-Coupled Matrix-Assisted Laser Desorption/Ionization-Direct Organelle Mass Spectrometry: A Technique for the Detailed Analysis of Single Organelles. J Am Soc Mass Spectrom 27:187–193
26. Huynh V, Joshi U, Leveille JM, Golden TD, Verbeck GF (2014) Nanomanipulation-coupled to nanospray mass spectrometry applied to document and ink analysis. Forensic Sci Int 242:150–156
27. Clemons K, Wiley R, Waverka K, Fox J, Dziekonski E, Verbeck GF (2013) Direct analyte-probed nanoextraction coupled to nanospray ionization-mass spectrometry of drug residues from latent fingerprints. J Forensic Sci 58:875–880
28. Ledbetter NL, Walton BL, Davila P, Hoffmann WD, Ernest RN, Verbeck GFIV (2010) Nanomanipulation-coupled nanospray mass spectrometry applied to the extraction and analysis of trace analytes found on fibers. J Forensic Sci 55:1218–1221
29. Clemons K, Dake J, Sisco E, GFT V (2013) Trace analysis of energetic materials via direct analyte-probed nanoextraction coupled to direct analysis in real time mass spectrometry. Forensic Sci Int 231:98–101
30. Lysy PA, Smets F, Sibille C, Najimi M, Sokal EM (2007) Human skin fibroblasts: from mesodermal to hepatocyte-like differentiation. Hepatology 46:1574–1585
31. Madsen L, Petersen RK, Kristiansen K (2005) Regulation of adipocyte differentiation and function by polyunsaturated fatty acids. Biochim Biophys Acta Mol basis Dis 1740:266–286
32. Ntambi JM, Kim Y-C (2000) Adipocyte differentiation and gene expression. J Nutr 130:3122S–3126S
33. Bruins AP (1991) Mass spectrometry with ion sources operating at atmospheric pressure. Mass Spectrom Rev 10:53–77
34. Cazes J (2004) Analytical instrumentation handbook. CRC Press, Boca Raton

Chapter 4

Development of Pico-ESI-MS for Single-Cell Metabolomics Analysis

Zhenwei Wei, Xiaochao Zhang, Xingyu Si, Xiaoyun Gong, Sichun Zhang, and Xinrong Zhang

Abstract

In this chapter, we introduced a Pico-ESI strategy for metabolomics analysis with picoliter-level samples. This Pico-ESI strategy was technically achieved by pulsed direct current electrospray ionization source (Pulsed-DC-ESI). This source could collect MS signals for a few minutes from a cell, enabling us to obtain large-scale MS^2 data of metabolite IDs in single-cell analysis. Further identification of the single-cell metabolome such as the database match and chemical modification to metabolome was thereby achieved. Technically, this source could ionize sample with no need of sample and electrode contact, which can be potentially applied for high-throughput analysis. We also introduced several strategies related to Pico-ESI to reduce the matrix interference especially for extremely small samples developed in our group, including step-voltage nanoelectrospray, picoliter sample desalting method, droplet-based microextraction method, and probe-ESI, etc. All these strategies have been successfully applied to single-cell analysis.

Key words Pico-ESI-MS, Single cell, Metabolomics, Matrix interference

1 Introduction

The large-scale study on systematical profiles of metabolites is very important for cell biology [1–3]. Till now the information obtained from the fingerprints of metabolites are the results of the stochastic average masked by bulk measurement [4–9]. It is well known that the study on single-cell metabolomics is important and new discoveries might be obtained with cellular heterogeneity [10–14]. However, the big challenge was that a large volume of samples are required to obtain the systematical profiles of metabolites in MS analysis while only picoliter sample is available from a single cell. Therefore, the development of sensitive, high-throughput, and matrix-free MS methods to meet the requirement of picoliter sample analysis is urgently demanded.

In recent years, several groups have contributed their novel MS methodologies to single-cell analysis. For instance, Sweedler et al.

Bindesh Shrestha (ed.), *Single Cell Metabolism: Methods and Protocols*, Methods in Molecular Biology, vol. 2064,
https://doi.org/10.1007/978-1-4939-9831-9_4, © Springer Science+Business Media, LLC, part of Springer Nature 2020

utilized matrix-assisted laser desorption ionization [15] (MALDI) and capillary electrophoresis nanoelectrospray [16] (CE-Nano-ESI) MS for single-cell analysis of giant neurons of mollusks, leading to the discovery of new neuropeptides. The video mass spectrometry [17, 18] developed by Masujima et al. made it possible to sample single-cell extract into nanoelectrospray (Nano-ESI) emitter followed by mass spectrometry (MS) detection and principal component analysis (PCA). Zenobi et al. further studied the metabolites heterogeneity of different individual cells using MALDI-SCMS with high-throughput [11, 13]. Vertes et al. performed metabolites analysis at the subcellular level [19] by laser ablation ESI-MS (LAESI-MS). These studies have opened a new frontier in MS field and would lead to big progress in single-cell biology.

In our opinion, the next step of SCMS study may be on "omics" scale, such as metabolomics at the single-cell level. To achieve this goal, some technical problems should be solved: (1) Single cell is a typical sample with only 1 pL to 1 nL. Achieving relatively long spray time of the limited sample is an important issue because it at least takes a few minutes to obtain the data for the minimum requirement of metabolome identification. (2) The high-throughput SCMS analysis would be potentially applied to drug screening, immunology, and oncology. The obstacle is how to ionize the single-cell-based droplet arrays automatically. The development of a sample-electrode contactless Pico-ESI source would help to the realization of high-throughput analysis. (3) Single-cell samples contain complicated components and matrix. The development of SCMS methods for matrix removal and high-sensitive detection would be an important issue.

In this chapter, we would like to introduce a Pico-ESI strategy for metabolomics analysis with picoliter-level samples. This Pico-ESI strategy was technically achieved by pulsed direct current electrospray ionization source (Pulsed-DC-ESI). This source provided several minutes long MS signal from a single cell. These advantages enabled us to obtain large-scale MS^2 data of metabolite IDs from a single cell. Further identification of the single-cell metabolome such as the database match and chemical modification to metabolome was thereby achieved. Technically this source could ionize sample solution without sample-electrode contact, which could be potentially applied for high-throughput analysis. To achieve real animal/plant cell analysis, we also introduced several strategies developed in our group to reduce the matrix interference especially for extremely small volume of samples, including step-voltage nanoelectrospray, picoliter sample desalting method, droplet-based microextraction method, and probe-ESI.

2 Materials and Instrument Settings

2.1 Mammalian Cell Culture

2.1.1 Materials and Equipment

MCF-7 cells were purchased from ABGENT. The culture medium, trypsin-EDTA, PBS, 6-cm cell culture dishes and other materials used in cell culturing were all sterile and purchased from Corning (NY, USA).

All the cells were incubated in Sanyo CO_2 incubator (MCO-15AC). All the manipulations concerning mammalian cell culture were conducted in a laminar flow cabinet (Suzhou Zhijing Purification Equipment Co., Ltd., SW-CJ-2FD) to prevent micro-organisms contamination. Before conducting any operation in the cabinet, the UV- germicidal lamp must be on for at least 30 min. The UV light must be turned off during use to avoid the exposure of skin and eyes to ultraviolet light. In addition, the laminar flow must be switched on during use. All the reagents and materials must be sanitized with 70% ethanol in water before being placed into the cabinet.

2.1.2 Preparing Culture Medium

MCF-7 cells were grown in Dulbecco's minimum essential medium (DMEM) supplemented with 10% fetal bovine serum (FBS) and 1% penicillin/streptomycin in a humidified incubator containing 5% CO_2 at 37 °C. The culture medium had better be used up within 1 week after the medium was prepared.

2.1.3 Passaging Cells

As the cells proliferate in the medium, the number of cells in the culture dish grows. When the number of cells in a 6-cm culture dish was over 1×10^6, a small number of cells must be transferred into a new dish with fresh medium added to culture the cells for a longer time. The procedure to passage cells includes:

1. Warm the medium at 37 °C.
2. Rinse the cell monolayer with spent culture medium to suspend the dead cells in the dish. Since the live MCF-7 cells would attach to the bottom of the dish, while the dead cells suspended in the medium, by rinsing the cells, all the dead cells would suspend in the medium. Remove the spent medium and all the dead cells from the dish and discard.
3. Add 2 mL PBS into the dish. Rinse the cell monolayer with PBS. Remove PBS and discard. Repeat the step twice to remove residual medium in the dish completely, because the serum in the medium contains trypsin inhibitor which would prevent the tryptic digestion in the next step.
4. To harvest the cells, 1 mL trpsin/EDTA was added into the dish to detach the Hela or MCF7 cells. After 3-min incubation at room temperature, remove the trpsin/EDTA and discard. At this moment, the cells are already detached, so the trpsin/

EDTA must be aspirated very carefully to avoid disturbing the cell monolayer.

5. Add 1 mL fresh medium into the dish and pipette the cells up and down to disperse the cells into a single-cell suspension. Add appropriate volume of cell suspension (usually 200–500 μL) into a new dish. Add 4 mL fresh medium into the new dish. Gently rock the dish back and forth to disperse the cells in the medium.
6. Place the dish in the incubator until passaging the cells again 2 or 3 days later.

2.2 Prepare the Emitter for Pico-ESI

The borosilicate glass capillaries (I.D. = 0.6 mm, O.D. = 1 mm, without filament) were purchased from Vital Sense Scientific Instruments Co. Ltd. Borosilicate glass capillaries were pulled by P-2000 (Sutter Instrument) to make the emitters (I.D. of the tip is 1 μm, I.D. = 0.6 mm, O.D. = 1 mm, length = 55 mm). The parameters of P-2000 were as followed: Heat = 345, FIL = 5, VEL = 28, DEL = 128, PUL = 60.

2.3 Setting the Single-Cell Manipulator System

All the single-cell sampling experiments were performed with the aid of three-dimensional mobile manipulator (MP-225, Sutter Instrument). The procedure was observed by an inverted microscope (D × 30, Dayueweijia Science and Technology Co. Ltd., Beijing). Typically, six was chosen as the microstep level for experiment.

2.4 The Mass Spectrometry Parameters

Accurate mass measurements were accomplished on Orbitrap MS (Q-Exactive, Thermo Scientific, San Jose, CA). Capillary temperature: 320 °C, tube lens voltage: 50 V, mass resolution: 70,000, maximum inject time: 50 ms, and microscans is 1. Other MS experiments were accomplished on LTQ MS (Thermo Scientific, San Jose, CA). Capillary temperature: 275 °C, capillary voltage: 9 V, tube lens voltage: 100 V, maximum inject time: 200 ms, and microscans is 1. The commercial ionization source of ESI was removed ahead of our experiments.

2.5 Metabolite IDs Identification in Pico-ESI-SCMS

In order to obtain detailed structure information of metabolite IDs, we used full MS—ddMS^2 analysis method (Set in the Xcalibur software, QE-Orbitrap). The detailed settings were as followed. Method duration was 3 min (for Allium cepa cell) or 5 min (HeLa cell). Full MS: resolution = 70,000, AGC = 3e6, max injection time = 100 ms, scan range = 134–2000 (Allium cepa cell) or 68–1000 (HeLa cell). The ddMS2 settings: resolution = 35,000, AGC = 1e5, ion injection time = 100 ms, loop count = 10, NCE = 50, underfill ratio = 0.1%, dynamic exclusion = 300 s.

3 Methods

3.1 Pulsed Direct Current Electrospray Ionization Source [20]

3.1.1 Method

In order to achieve Pico-ESI, which could elongate the sample spray time for large-scale metabolome data collection of single-cell sample, we had developed pulsed direct current electrospray (Pulsed-DC-ESI) source. Nano-ESI generated continuous electrospray while the mass analyzer of MS worked intermittently. Therefore, if we could generate pulsed electrospray and perfectly synchronized its frequency with the mass analyzer, we could achieve Pico-ESI without loss of MS sensitivity by avoiding meaningless sample loss between the time-adjacent MS scan events.

Figure 1a shows the setup of Pulsed-DC-ESI. After the sample was loaded into the emitter, the emitter was placed in a holder for Pulsed-DC-ESI analysis. The holder included a straight stainless steel tube (I.D. = 1.2 mm, length = 3 cm). The rear part of the emitter was inserted into this tube thereby the emitter could be fixed. Inside the holder, a steel needle electrode (O.D. = 0.3 mm) was inserted into the emitter from its rear part. The electrode is contactless from the sample solution with a typical distance of 5 mm. DC voltage was applied to this needle to provide the static electrical field for electrospray. The holder was fixed on the cantilever of a four-dimensional mobile device (x, y, z, and θ_y, Beijing Optical Century Instrument Co., Ltd). By adjusting the mobile device, the distance between the emitter tip and MS inlet and spray angle could be controlled. Typically, the electrode voltage was +1.5 kV for positive mode analysis and −1.2 kV for negative mode analysis; the frequency of the pulsed spray could be controlled by electrode voltage (Fig. 1b). The polarity of electrospray could be controlled by the polarity of DC power supply (Fig. 1c).

3.1.2 Results

Almost 100% Sample Utilization Ratio Without Sample Dilution

For most of the lab doing single-cell analysis, the emitter pulled by the commercial instrument (P-2000, Sutter Instrument) was widely used because of the economy and the well-controlled orifice diameter (10 nm ~ 10 um). However, the electrical contact for these emitters appears a serious problem. As indicated by "*In ESI, electrical contact with a voltage supply is necessary to generate a continuous spray of charged droplets from a solution. The electrical contact adds dead volume and adsorption surfaces*". This is a critical problem when performing analysis of small volume sample such as single cell. In our previous study, this problem was solved by adding the spray solvent into the emitter to help electrical contact of the electrode and sample solution [17]. While there is an obvious drawback that the sample could be diluted by the spray solution. In Pulsed-DC-ESI, we utilize DC power supply to build up the static electrical field to help the pulsed ionization of sample solution. This is convenient for the wide single-cell analysis researchers.

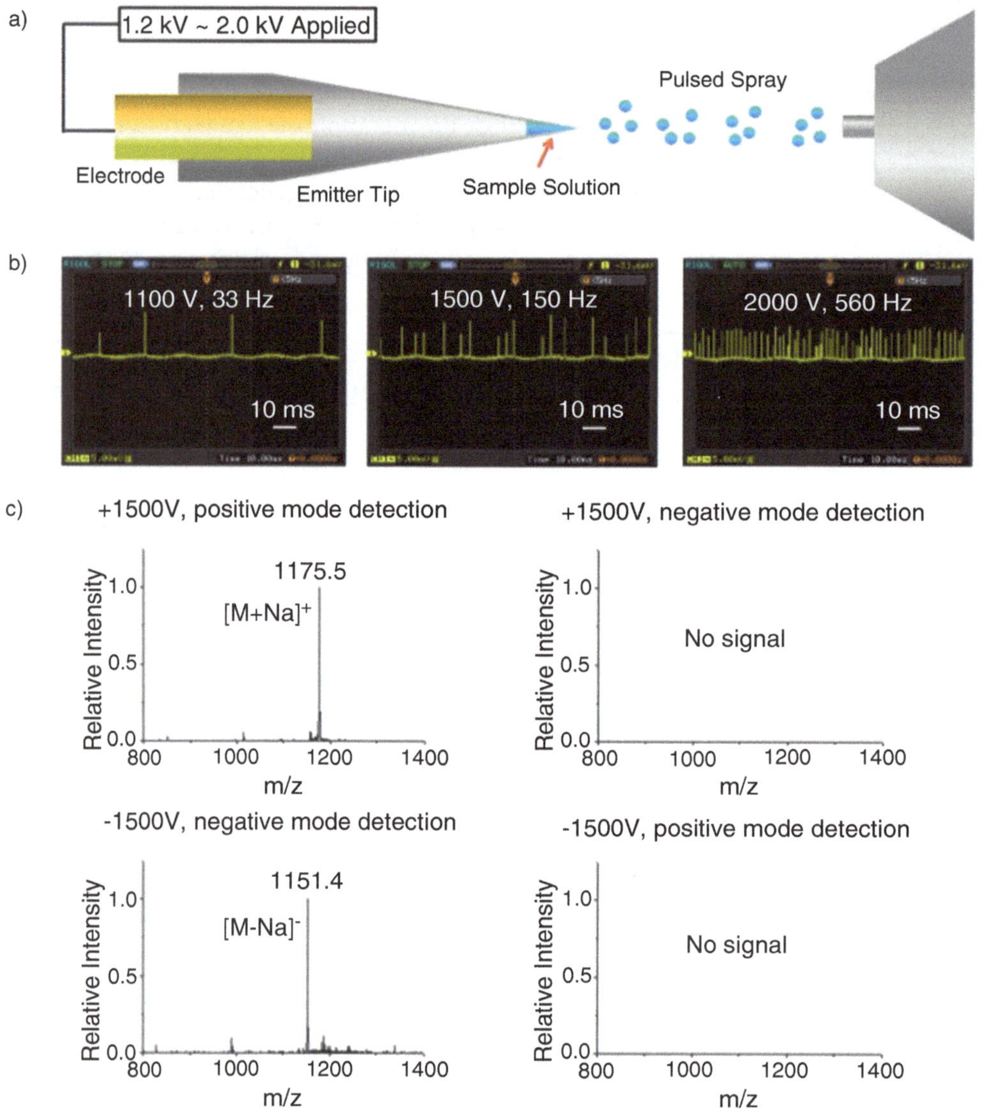

Fig. 1 (**a**) The setup of Pulsed-DC-ESI. (**b**) Adjusting the pulsed spray frequency by electrode voltage. 1 μL Somatostatin (10 ppm) was loaded into the emitter. The probe of oscilloscope was set very near to the emitter tip to detect the pulsed electrospray. (**c**) To control the polarity of Pulsed-DC-ESI by the polarity of power supply. The testing sample was 100 ppm maltoheptaose

Picoliter Per Minute Sample Flow Rate

The sample flow rate is very important when analyzing a small volume sample with complicated components. Lower sample flow rate helps to elongate the sample spray time, providing the time to obtain more information from the volume-limited sample. Pulsed-DC-ESI achieve ultra-low sample flow rate (~150 pL/min) due to the pulsed sampling strategy. By comparison to Nano-ESI when +1.5 kV Dc voltage applied, the flow rate of Pulsed-DC-ESI reduced to 1.9% ~ 4.2% of Nano-ESI (Table 1). Take the single-cell

Table 1
The average flow rate achieved by Nano-ESI and Pulsed-DC-ESI for different samples

Compounds	Polarity	Volume (pL)	Signal duration(s) Nano-ESI	Pulsed-DC-ESI	Pulsed-DC-ESI/Nano-ESI	Average flow rate Nano-ESI (nL/min)	Pulsed-DC-ESI (pL/min)	Pulsed-DC-ESI/Nano-ESI (%)
Indole acetic acid	Negative	370 ± 30	0.9	35.0	38.9	26.7	685.7	2.6
Jasmonic acid	Negative	370 ± 30	0.6	25.2	42.0	40.0	952.4	2.4
Somatostatin	Positive	370 ± 30	1.9	62.7	33.0	12.6	382.8	3.0
Angiotensin II	Positive	370 ± 30	0.7	31.1	44.4	34.3	771.7	2.3
Maltoheptaose	Positive	370 ± 30	1.2	57.3	47.8	20.0	418.8	2.1
Cytochrome C	Positive	370 ± 30	1.2	28.6	23.8	20.0	839.2	4.2
Allium cepa cell extract	Positive	~ 900	6.6	345.1	52.3	8.2	156.5	1.9

analysis of *Allium cepa* as an example, the 150 pL/min sample flow rate enabled us to obtain the continuous MS signal of extract from a single sample for about 3 min. During the 3 min, we could obtain MS^2 more than 1000 components in the sample (by QE-Orbitrap), which would be of great help to the identification of metabolite IDs.

Single-Cell Metabolomics Analysis

To identify the metabolite IDs, the MS^2 data is very important. Full scan data-dependent MS^2 strategy (dd-MS^2) is a powerful MS analytical tool for this task. In dd-MS^2 analysis, a sequence of scan event of full scans and MS^2 scans is designed. The data dependent means the mass number of parent ions of MS^2 scans are dependent to the mass number appeared in the full scans. For example, we had profiled 1034 components from an individual *Allium cepa* cell (Fig. 2a) and 656 components from an individual HeLa cell (Fig. 2b) with Pico-ESI strategy. The long signal duration was essential to the acquisition of such "big data", making it possible to collect abundant MS^2 data and build the 2-D mass spectrum (including exact m/z number and MS^2 information) of metabolome. The most outstanding advantage of the "big data"-based 2-D mass spectrum was that it could give more information of low abundant metabolites, even the metabolites submerged in the noise of full scan. For example, we could barely figure out the low abundant metabolites components in the m/z range of 1000–2000 of *Allium cepa* cell since many of them were submerged in the noise. However, these components could be clearly visualized in

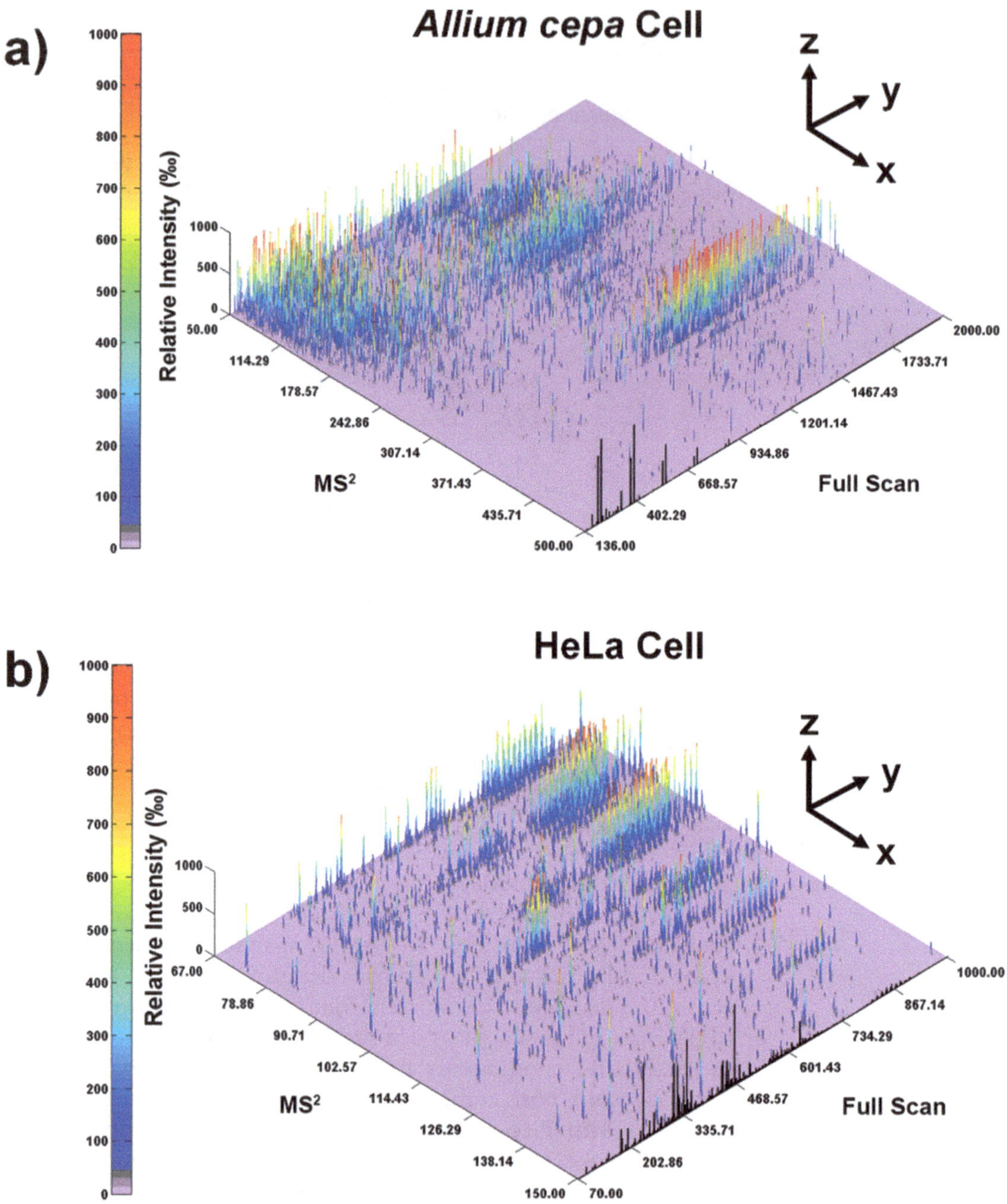

Fig. 2 The profile of metabolites from an individual (**a**) *Allium cepa* and (**b**) HeLa cell by Pico-ESI strategy. The massive MS^2 data of the 1034 components and their accurate *m*/*z* values were used to build a 2-D mass spectrum. The full scan axis (*y* axis) indicated the *m*/*z* of the parent ions. At a certain *m*/*z* of full scan axis, the color sticks along the MS^2 axis (*x* axis) showed the relative intensity of MS^2 fragments of the corresponding *m*/*z*

the 2-D mass spectrum (Fig. 2). With the abundant MS^2 data, we could perform small omics scale analysis, such as database search, typical fragments search, and neutral loss search to identify the single-cell metabolome more accurately.

3.2 Step-Voltage Nanoelectrospray [21]

3.2.1 Methods

Biological samples are very complicated, and the matrices, especially the containing salts, have a significant ion suppression effect, which enormously reduces the efficiency of ionization. Moreover, the salts in samples can form complicated compounds through molecular association and coordination, which makes it quite difficult to well identify the target compounds in mass spectrum. It is even worse when using Nano-ESI-MS for analysis of single cell due to its very small sample volume. Although the separation methods, such as liquid chromatograph (LC), capillary electrophoresis (CE), and solid phase micro-extraction (SPME), have been frequently applied for salt removal in order to reduce the matrices interference, the process is relatively tedious. Moreover, the separation procedures for LC or CE may cause a sample dilution during elution especially for ultra-small volume of samples.

Step-voltage nanoelectrospray ionization method (SV-Nano-ESI) was thereby designed to remove the matrices interference when using Nano-ESI for small volume sample analysis (Fig. 3). The step voltage is shown in Fig. 3b. A voltage of 5.2 kV was first applied on the solution for 30 s. A current of 10 μA was recorded. During this period, separation of analytes and matrices happened due to ion electromigration in the tip zone of the emitter. Subsequent nanoelectrospray was carried out at a voltage of 2.4 kV for the sample analysis. Under the action of strong electric field, salt cations and anions in matrices would rapidly migrate into the opposite directions in the tip zone, forming the leading band and tailing band. Due to the low electromigration rates, the target proteins maintained almost in the middle band and could be analyzed without the matrices interference and high signal-to-noise ratio was achieved.

3.2.2 Results

Possible Mechanism of SV-Nano-ESI

The possible mechanism of matrices removal in the microzone of the emitter tip could be the enhanced electromigration, which separates analytes and salts with different electromigration rate. Ions electromigration and solution migration are two important processes during ESI. Taking positive voltage mode as an example, ions electromigration causes cations migrate to the tip of emitter and anions migrate to the anode. The solution migration, which is forced by the formation of jet flow at the tip, causes continuous solution consumption. Different from Nano-ESI, during SV-Nano-ESI, ions electromigration is much faster than solution migration. As a result, the separation of ions with different

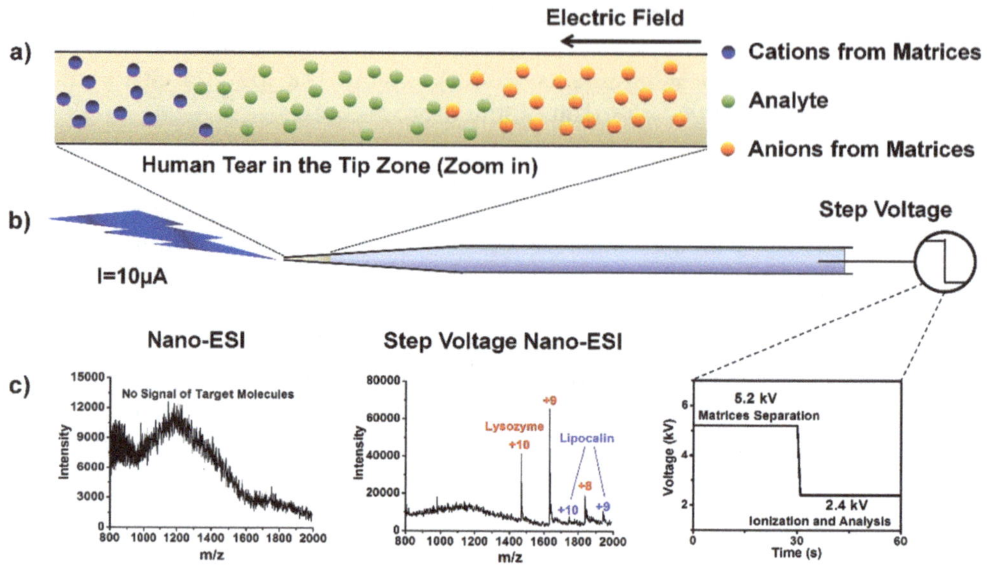

Fig. 3 (**a**) The ion electromigration for separation of matrices and analyte. Respectively, the salt cations and anions in the matrices migrated to the leading band and tailing band and the analyte kept still in the middle band due to their different electromigration speed. (**b**) The schematic of SV-Nano-ESI. The emitter (orifice = 15–20 μm) was loaded with a sample solution by a micropipette. A metal wire conductor was then inserted into the capillary. Step voltage was applied to the emitter. At the first 30 s, a current of 10 μA was recorded, indicating the action of ions under the strong electric field. Subsequently, a +2400 V voltage was applied for electrospray and MS analysis in a positive ion mode. (**c**) The mass spectra of human tear sample by using conventional nanoelectrospray and SV-Nano-ESI, the matrices interferences is removed apparently

electromigration rate could be observed. For example, we analyzed a sample of 10 ppm angiotensin II and 100 ppm maltoheptaose in Tris-EDTA buffer solution (Fig. 4). While SV-Nano-ESI, inorganic ions such as metal ions, ligand anions, and their coordination compounds carried high apparent charge(s) and migrated fast to zone 1 and zone 4 oppositely with a delay time of 2 s and 42 s respectively, left a 40 s long zone without matrices interference. The biological analytes such as angiotensin II and maltoheptaose carried low apparent charge(s) in the solution. They preferred to distribute in zone 2 and 3. Therefore, the molecules with different apparent charge(s) were separated from zone 1 to 4 after discharge (Fig. 4b).

Application of SV-Nano-ESI for Small Volume Sample Analysis

SV-Nano-ESI has been successfully used in the analysis of a small volume sample with complicated matrices (Fig. 5). For instance, when using Nano-ESI, the MS signal of angiotensin II, maltoheptaose, and lysozyme submerge in the interference signals of Tris-EDTA buffer solution. However, when SV-Nano-ESI was applied,

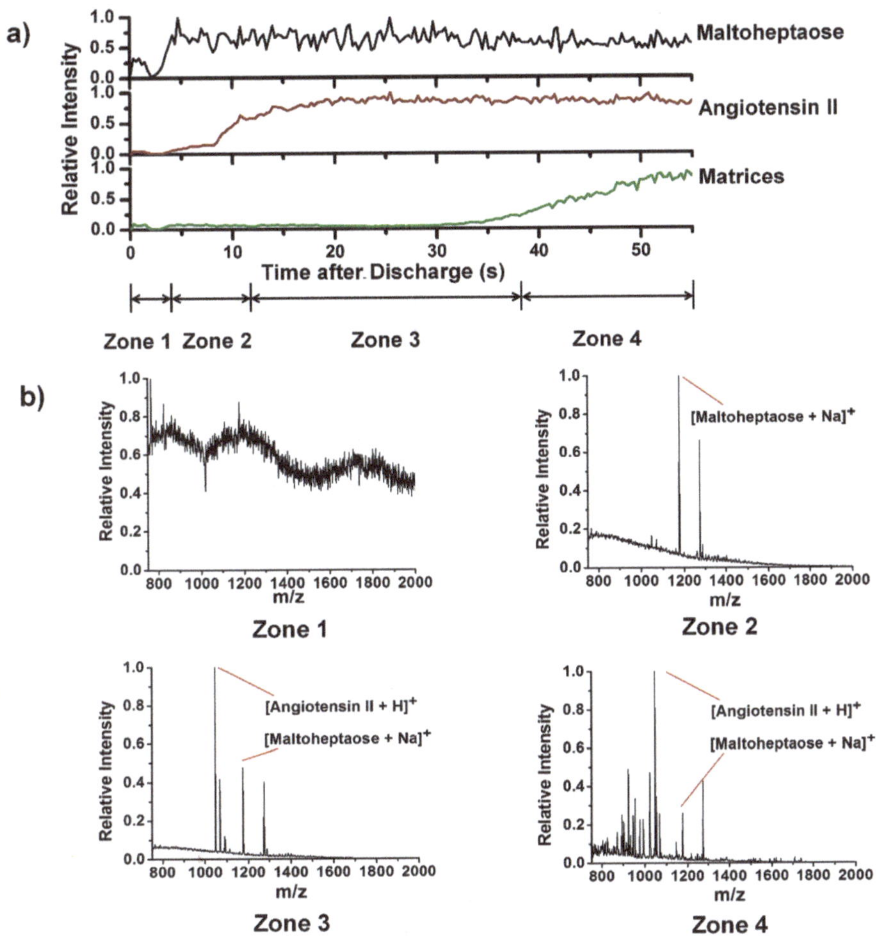

Fig. 4 (**a**) The extracted ion chronogram of different ions in the tip zone after high voltage applied in positive mode. The pH of the solution is nine. The extremely high electric field forced different ions to migrate, salt cations and anions from the matrices migrated the fastest and distributed more in zone 1 and zone 4; maltoheptaose and angiotensin II migrated slow and distributed more in the zone 2 and zone 3. (**b**) The mass spectra of zone 1 to zone 4. The ratio of intensity of angiotensin II and maltoheptaose in zone 3 was chosen to evaluate the stability. The RSD is 2.3% (under 7 times analysis)

the spectra became clean and the interference signal of matrices are removed. It is well known that urine and tear are both very complex biological samples, which are hard to be analyzed by nanoelectrospray without any pre-treatment prior to analysis. However, adopting SV-Nano-ESI for analysis gave a 4.3 of SNR for angiotensin II in urine sample, and a 6.5 of SNR for lysozyme in human tear. These results suggest that the developed SV-Nano-ESI plays an important role in improving the sensitivity of the target compounds in small volume biological samples such as physiological fluids and single cell.

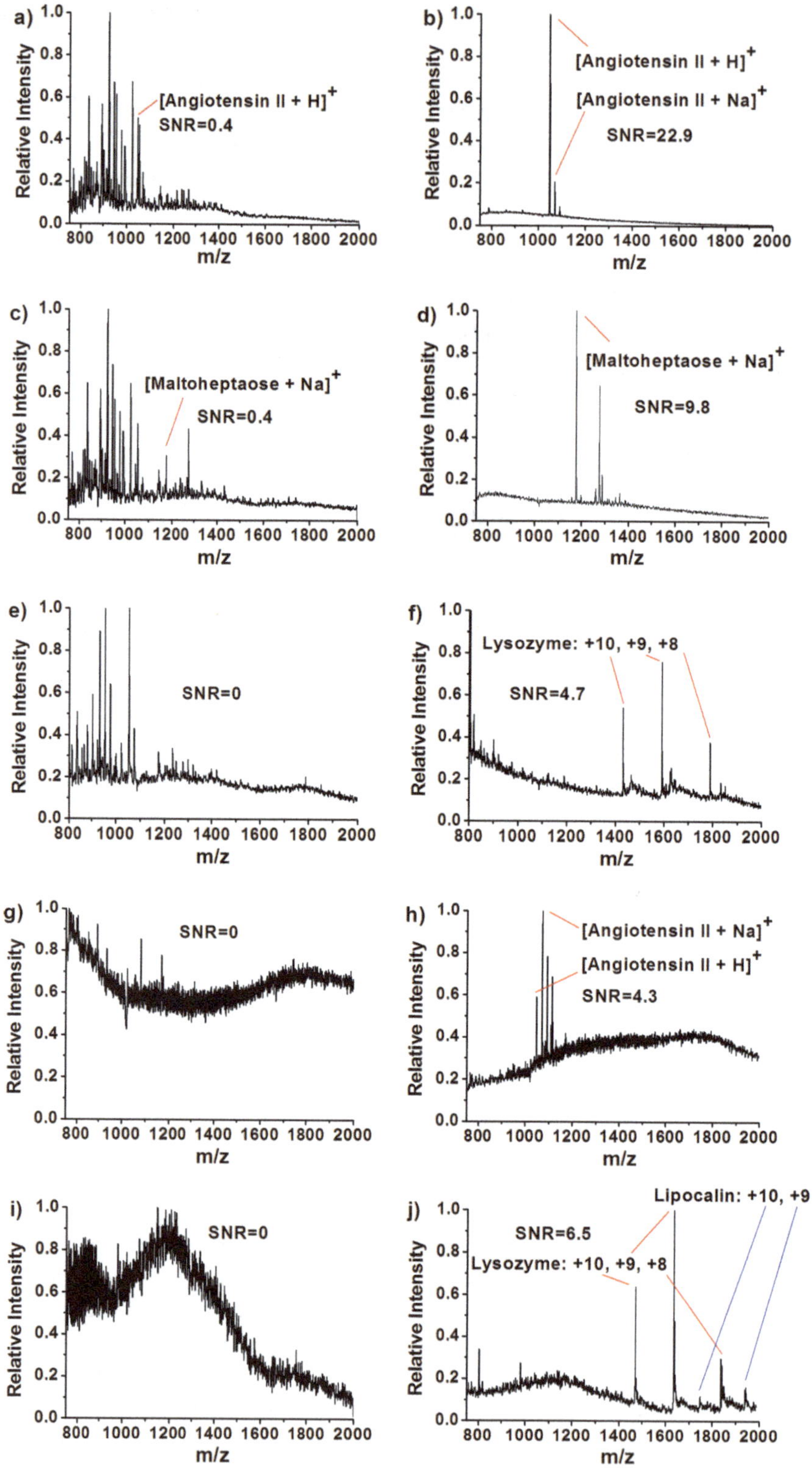

Fig. 5 The performance of Nano-ESI (**a**, **c**, **e**, **g**, **i**) and SV-Nano-ESI (**b**, **d**, **f**, **h**, **j**) in analysis of different samples. Samples of a to f are made in Tris buffer solution: 10 ppm angiotensin (**a**, **b**), 100 ppm maltoheptaose (**c**, **d**), 100 ppm lysozyme (**e**, **f**). Sample of **g** and **h** were made by adding 10 ppm angiotensin II into human urine. (**i**) and (**j**) are human tear sample. Each time, <100 nL sample was loaded for analysis and 10 points average was used to acquire the mass spectrum

3.3 Desalting Pico-Liter Samples by Crystallization

3.3.1 Methods

Biological researches often involve analysis of samples with extremely small volumes. The successful detection of biomolecules in small volume samples is still challenging in many cases due to the high concentrations of salts added in the solutions during pretreatment. These salts are used to maintain the pH value and osmotic pressure of the aqueous solutions so that cells could keep alive. Salts are also added in solutions to maintain the structure and biological activities of proteins. However, even low milli-molar concentrations of some metal ion salts can cause severe ion suppression and peak broadening due to the formation of clusters and adducts. Here, we introduce a method namely desalting by crystallization for salts removal from picoliter buffer solution to enhance the Pico-ESI-MS sensitivity for small volume sample analysis.

Sampling

Inserting the emitter into insulin solution, then the solution was sucked into the emitter's tip via spontaneous capillarity. The inner diameter of the tip was 1–2 μm. The sampling volume was controlled by fixing the sampling time, which was around 1 s, and the sampling volume was typically 20–100 pL. For nano-ESI, the sampling volume was controlled as 3–5 μL to fill the emitters.

Desalting (Crystallization)

After sampling, the emitters were placed in the open air at room temperature (20–25 °C) to evaporate the solvents naturally (Fig. 6a). A lower air humidity would facilitate the desalting process. The desalting process usually would only take a few minutes. The process could be faster if the samples contained a certain proportion of volatile solvents.

Pico-ESI Analysis

1. After desalting, the emitter was placed in front of the MS inlet with a distance of 5 mm. A copper wire was inserted into it until the wire reached near the emitter's tip. The distance between the wire's tip and the emitter's tip (Fig. 6b) was about 1.5 mm.
2. A high D.C. voltage of 1750 V (+1750 V for positive mode, and −1750 V for negative mode) was applied to the copper wire.
3. A droplet of assistant solvent (methanol/ water = 1:1, with 0.1% HAc) was used to bedew the emitter's tip, which aimed to wet the biological sample in the tip (Fig. 6b). The volume of the solvent droplet was controlled using a pipette. The droplet was usually 15 μL and was kept at the tip for 1 s. Then the droplet was moved away and the picoliter solution in the emitter was for subsequent Pico-ESI analysis (Fig. 6c).

3.3.2 Results

Desalting and Detection Capability

The desalting and detection capability of this method was demonstrated by increasing the concentrations of the salts in insulin solutions (Fig. 7). The concentration of insulin solution was kept at 500 nM, while NaCl was ranged from 0 to 150 mM (The concentration of physiological saline). Videlicet, the max concentration of

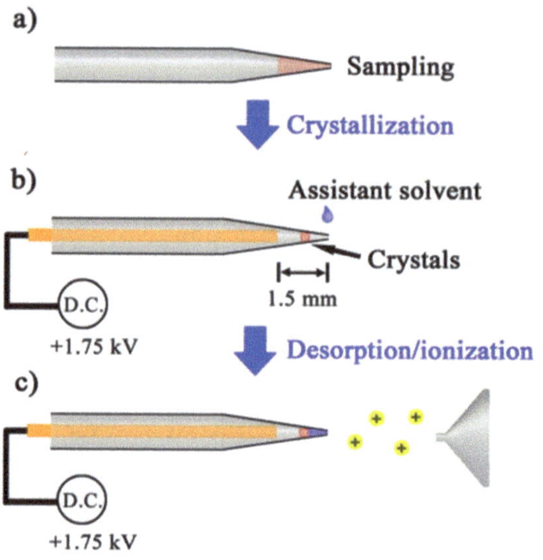

Fig. 6 Schematic setup of the method. (**a**) Sampling. Samples were sucked into the tip via spontaneous capillarity. (**b**) Wetting of the tip. A droplet of assistant solvent was placed at the tip for a fixed time to wet it. (**c**) Desorption/ionization process. After wetting, the biomolecules were dissolved into the assistant solvent and immediately motivated by the electric field to spray out from the tip into MS

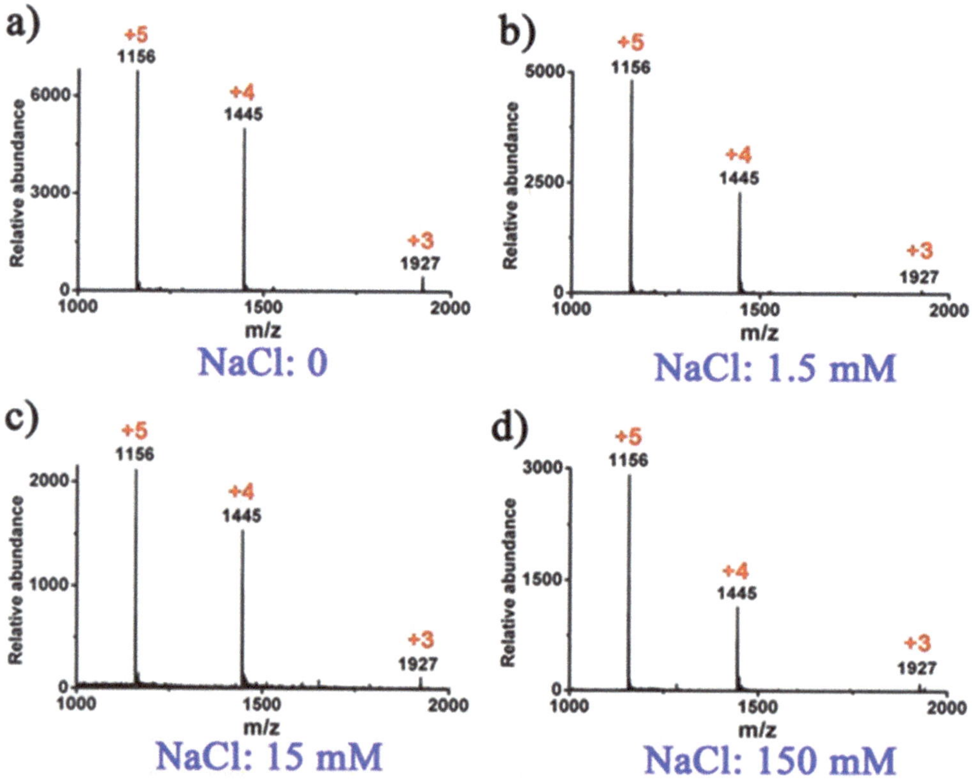

Fig. 7 Detection results of insulin aqueous solutions with different concentrations of NaCl. The mass spectra were obtained in positive mode. (**a**)–(**d**) The concentrations of NaCl were 0, 1.5, 15, and 150 mM

NaCl was 3×10^5 folds more than that of insulin in the samples. The sampling volume was controlled as 100 pL for each time, corresponding to 50 amol insulin. As the obtained mass spectra in Fig. 7 shows, insulin could be successfully detected regardless of the increasing NaCl.

Mechanism of Desalting

To elucidate the desalting mechanism, an aqueous solution of 500 nM gentian violet was used as the sample (Fig. 8). Additional 1.5 mM NaCl was added into the sample. The sampling volume was 30 pL, as was calculated with a microscope (Fig. 8a). Thus the sampled gentian violet in the pipette was 15 amol. After sampling, the sample was left dry under room temperature to allow the salt crystallize. Afterwards, the separated gentian violet was directly desorbed/ionized from the tip of the pipette using Pulsed-DC-ESI for MS detection. As was shown in Fig. 8f, the 15 amol gentian violet was successfully detected using our method. In comparison, conventional nano-ESI failed in the detection.

In the desorption/ionization process, the organic assistant solvent (Acetonitrile) quickly dissolved the gentian violet molecules surrounding the NaCl crystals. The dissolved gentian violet

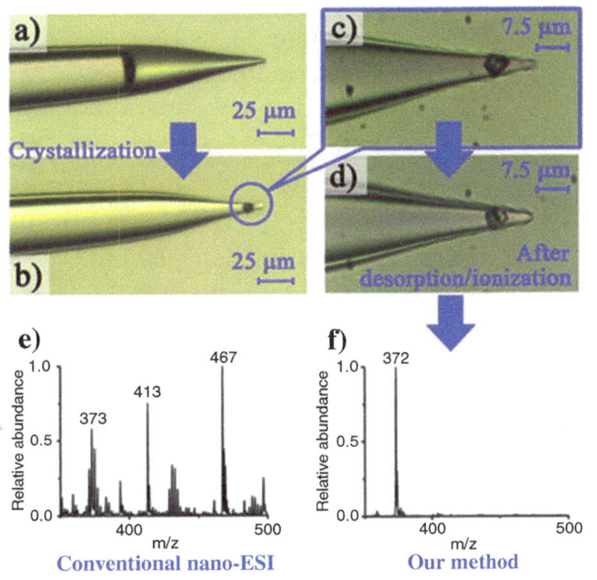

Fig. 8 Desalting and desorption/ionization of gentian violet investigated by microscope. (**a**) Gentian violet solution sucked in the tip. (**b**) NaCl crystal formed in the tip after crystallization. (**c**) Amplified image of the NaCl crystal formed in the tip. (**d**) NaCl crystals remained in the tip after the desorption/ionization procedure. The assistant solvent used for desorption/ionization was pure acetonitrile. (**e**) The obtained mass spectrum using conventional nano-ESI in positive mode. (**f**) The obtained mass spectrum using our method in positive mode. Gentian violet was represented by the peak of $m/z = 372$, losing a Cl- during ionization

molecules were then motivated by the electric field to spray out from the tip and detected by MS. After detection, the NaCl crystals were checked again (Fig. 8d). Regardless of the desorption/ionization procedure of gentian violet, the NaCl crystals remained almost unchanged (compare Fig. 8c, d). This confirmed that gentian violet was actually separated from the salts during crystallization, and could be desorbed/ionized individually without salt.

3.4 Droplet-Based Microextraction for Detection of Metabolites in Mammalian Cells

3.4.1 Methods

The formation and development of many diseases are associated with specific metabolites in cells. It is significant to detect intracellular target metabolites at the single-cell level. However, matrix interference affects the detection. For instance, the cytoplasm of mammalian cell is usually multicomponent, there is interference between different components. Culture medium contains many molecules similar or identical to intracellular metabolites, which also has interference. Besides, buffer solutions with rich salts are usually used in the pretreatment of cell samples, which could cause severe ion suppression. Here, we introduce a method that integrates droplet-based microextraction and desalting by crystallization for removal of matrix interference followed by Pico-ESI-SCMS analysis.

Pretreatment of Cell

1. Before detection, cells in the petri dish were quickly washed with ammonium formate aqueous solution (150 mM, 4 °C) for three times. The ammonium formate solution was kept at 4 °C before use so as to stop the cellular reactions.
2. After wash, the ammonium formate solution was sucked away.
3. Then, the petri dish was placed in a vacuum drying oven for drying around 10 min.

Sampling

(1) After drying, the petri dish was placed under an inverted microscope for sampling. (2) Quartz emitter with a tip diameter of 2–3 μm was used. The very tip of the emitter was pre-filled with 1–3 nL solvent (Methanol/water = 1:3). (3) Then, a three-dimensional manipulator was used to precisely place the very tip of the emitter close to the surface of the cell. The solvent was pushed out from the tip onto the cell by a syringe. (4) The cell was then wrapped by the solvent for 10 s. Afterwards, the solvent was sucked back into the tip. (5) The obtained solution in the tip was let to dry spontaneously under room temperature, which would only take no more than 10 min. After the drying step, the extracted metabolites were enriched to the very tip of the emitter due to the gradual evaporation of the solvent. (6) Subsequent in situ desorption/ionization of the metabolites was achieved by the method mentioned in Subheading 3.3.1.3.

3.4.2 Results

Study of Unicellular ATP-ADP-AMP

The most well-known energy molecules, namely ATP-ADP-AMP, play vital roles in the energy metabolism of cells. The dynamic changes of ATP-ADP-AMP in individual cells before, during, and after division were studied, as shown in Fig. 9. According to the results, the energy states varied greatly among different individual cells. For the cell before division (Fig. 9a, b), ADP was the most abundant species. While for the cells during division, ATP dominated the spectrum (Fig. 9c, d). After division, the content of ATP dropped, and ADP turned to be the most abundant species (Fig. 9e, f).

Change of Concentration Ratios

We have made standard curves of ATP and ADP to calibrate the MS signal intensities into concentrations, and the concentration ratios of ADP/AMP and ATP/AMP in cells under different physiological states were obtained. For the cell before division, the ratio of ADP/AMP and ATP/AMP were roughly equal (Fig. 10a). However, for cells during division, the ratios of ADP/AMP and ATP/AMP were quite different. The ratio ATP/AMP was about four times higher than that of ADP/AMP (Fig. 10b). This might be caused by the exuberant metabolic activities during division, including the synthesis of materials and the stretch of microtubules, etc. A large quantity of energy was devoted to these activities, leading to a great demand for energy. So the synthesis of ATP was enhanced at these times. In comparison, for cells that had already finished division, the ratios of ADP/AMP and ATP/AMP returned to the original state (Fig. 10c). The detection results that obtained from a whole petri dish of cells were shown in Fig. 10d as a comparison.

3.5 Probe-ESI for Metabolites Analysis of Plant Cell at Unicellular and Subcellular Levels

3.5.1 Methods

Directly sampling a small amount of cell content into a nanospray tip for subsequent MS detection offers a way for live cell analysis. Till now, however, few methods had made use of an enrichment and separation procedure in single-cell analysis. Probe ESI (PESI) offered such an opportunity. Hiaroka proposed the concept of "probe electrospray ionization" and applied PESI in the analysis of several biological systems, such as plant and animal tissues [22, 23]. We have found that this technique was also suitable for unicellular and subcellular analysis: (1) direct sampling in live cells; (2) enrichment of metabolites, which improved the sensitivity; (3) subcellular detection could be achieved with miniaturized probes; (4) reduced disturbance to the original distributions of metabolites during sampling; (5) simplified device and reusable probes. It would be of great significance to introduce probe ESI into the field of single-cell analysis.

Sample Preparation

Purple *A. cepa* bulbs were used for experiments. They were stored at 4 °C before use. The *A. cepa* bulbs were washed with ultrapure water (Barnstead Nanopure, Thermo Scientific, San Jose, CA. Resistance $\geq$18 MΩ·cm^{-1}) to clean the surface of the epidermis. A monolayer of epidermal cells was directly removed from the bulb and mounted smoothly onto a clean microscope glass slide for subsequent experiments.

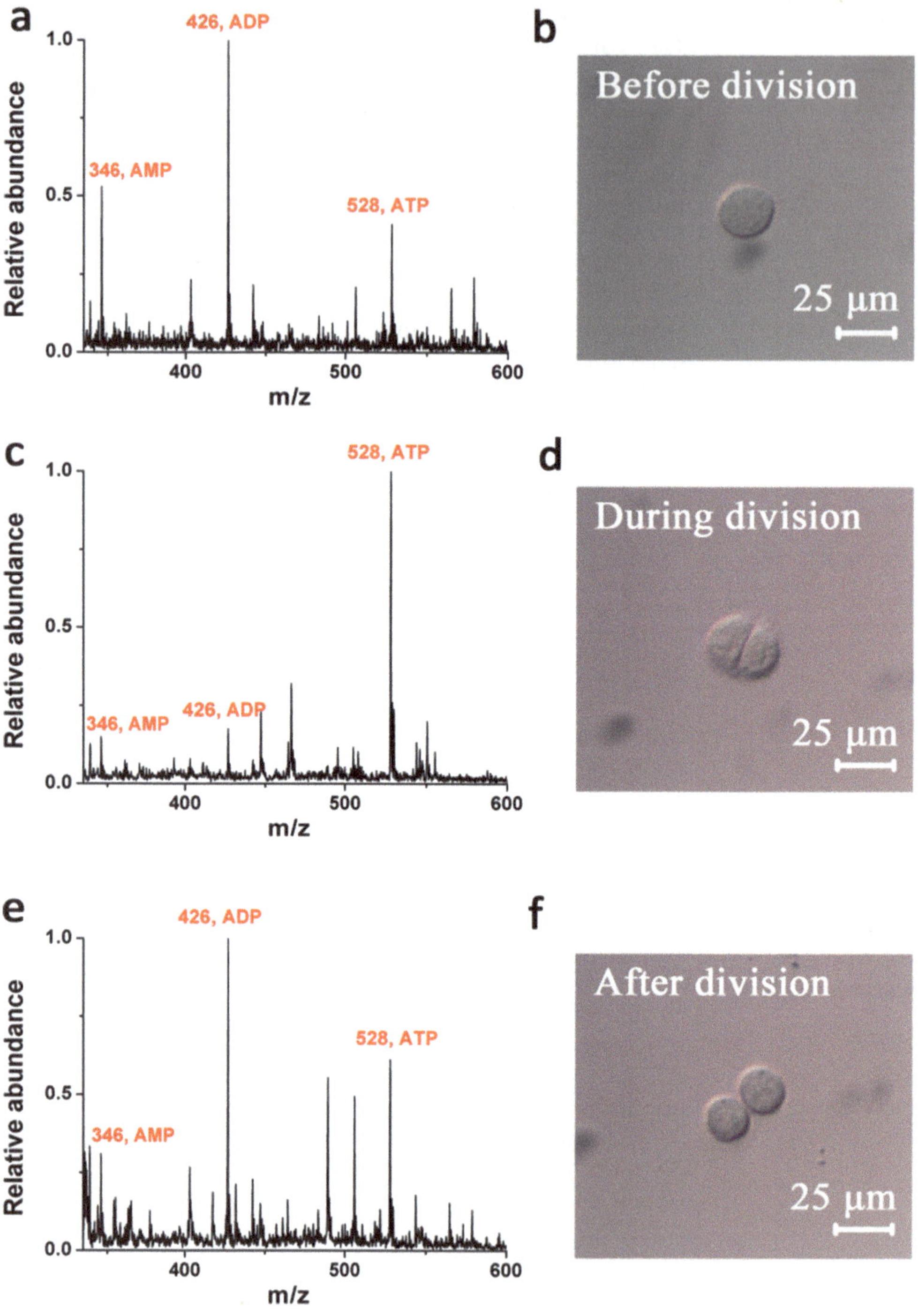

Fig. 9 The content changes of ATP, ADP, and AMP in MCF-7 cells before, during, and after division. The cells were passaged and cultured for 9 h before treatment. (**a**, **b**) The obtained mass spectrum and the corresponding image of the cell before division. (**c**, **d**) The obtained mass spectrum and the corresponding image of the cells during division. (**e**, **f**) The obtained mass spectrum and the corresponding image of the cells after division

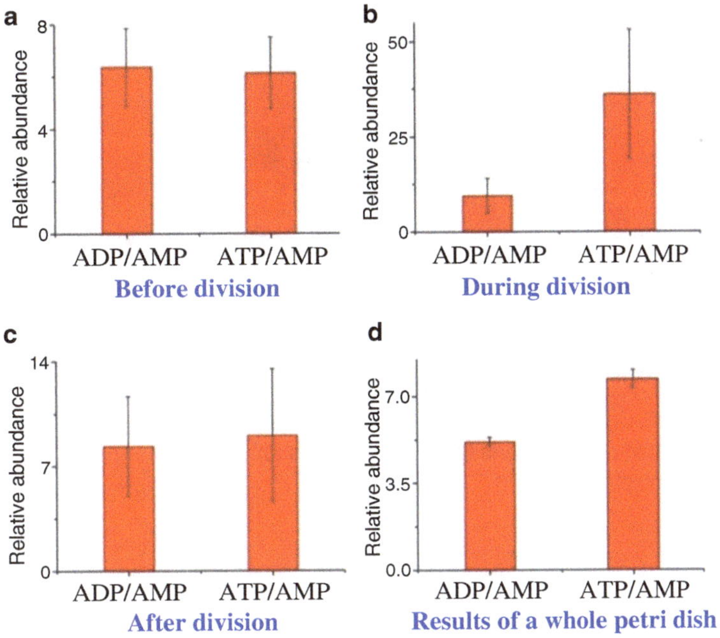

Fig. 10 The relative contents of ADP and ATP in cells under different physiological states. The cells had been passaged and cultured for 9 h before detection. (**a**) Cells before division. (**b**) Cells during division. (**c**) Cells after division. (**d**) The detection results obtained from a whole petri dish of cells

Probe Preparation

Tungsten probes with a tip diameter of 1 μm (99.9% tungsten, ST-20-5, purchased from GGB INDUSTRIES INC. USA) were used in the experiments. Before use, the probes were first washed with methanol/water (1:1) to clean its surface, and then soaked in 5% HNO_3 for 5 min. Afterwards, the probes were washed with ultrapure water, methanol/water (1:1), and ethanol, respectively.

Sampling and Detection

The probe was controlled by a three-dimensional manipulator (MP-225, Sutter Instrument Company, USA) to precisely insert into target cells (Fig. 11a). All the operations were observed with a microscope (D × 30, Dayueweijia Science and Technology Co. Ltd., Beijing). After insertion, the probe was kept in the cell for 30 s to enrich metabolites. Then, the probe was taken away from the cell and accurately positioned in front of the MS inlet with a distance of 5 mm for desorption/ionization (Fig. 11b). The tip of the probe was then wetted by sprayed assistant solvents, which was accompanied by undergoing desorption/ionization at applied high voltage (+2.5 kV for positive mode, and −2.6 kV for negative mode) for MS detection. Analytes that had been enriched on the tip were then desorbed/ionized from the tip and transferred into MS. A similar desorption/ionization method could be found in literature [22].

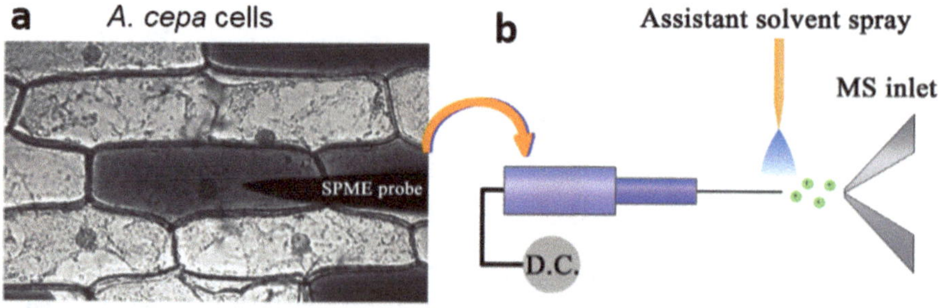

Fig. 11 Schematic of the probe ESI setup. (**a**) Microscopic image of the sampling procedure. Monolayer *A. cepa* cells were placed smoothly on a glass slice. The dark cells in the image were alive, with a certain amount of pigment in their vacuoles, while the light ones were dead. Only live cells were used in our experiment. (**b**) Setup of the coupling with MS. A D.C. high voltage (+2.5 kV for positive mode and −2.6 kV for negative mode) was applied to the probe. Assistant solvents were directly sprayed onto the tip of the probe to generate electrospray

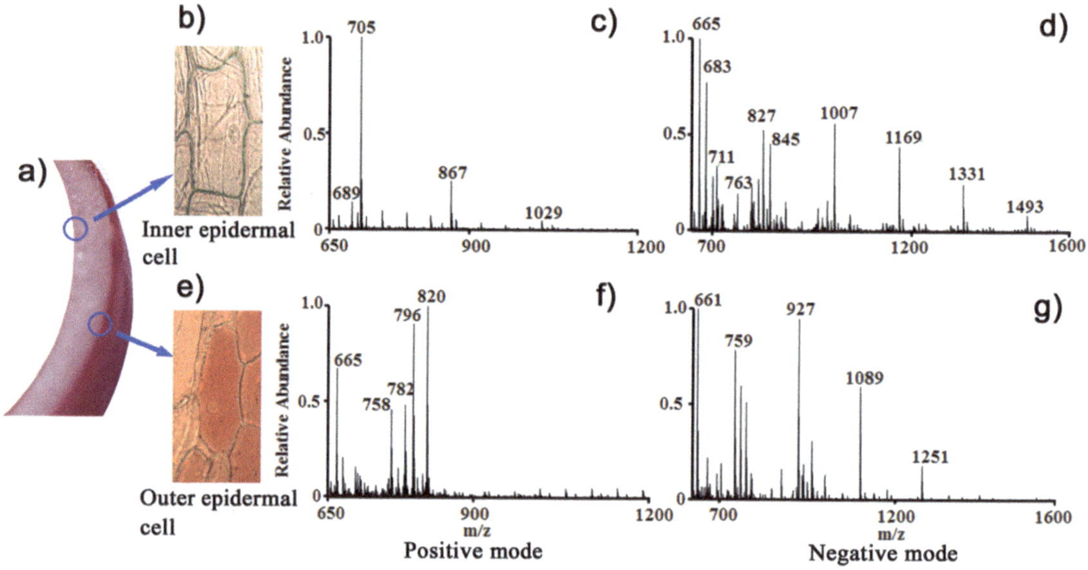

Fig. 12 Metabolite diversity of two different kinds of *A. cepa* cells revealed by their mass spectra. (**a**) Actual picture of an *A. cepa* bulb layer; Microscopic image (**b**) and the corresponding mass spectra (**c**, **d**) of an inner epidermal cell; Microscopic image (**e**) and the corresponding mass spectra (**f**, **g**) of an outer layer cell. The left row of mass spectra was in negative mode, and the right row was in positive mode. The enrichment time was 30 s and the assistant solvent was methanol/water/chloroform = 3:1:1

3.5.2 Results

Single-Cell Analysis

Two different kinds of cells, the outer epidermal cells and the inner epidermal cells, were analyzed at a cellular level (Fig. 12). Significant differences of metabolite species between the two kinds of cells have been observed. For example, six fructans (DP4-DP9, degree of polymerization) were assigned in inner epidermal cells (*m/z* 705, 867, 1029 in positive spectrum, and *m/z* 665/683,

827/845, 1007, 1169, 1331, 1493 in negative spectrum), while only three fructans (DP4-DP6) were assigned in outer epidermal cells (*m/z* 705, 867, and 1029), indicating that inner epidermal cells had more fructan species than outer epidermal cells. Although both inner and outer epidermal cells had fructan DP4-DP6, these three fructans seemed to be the major species in inner epidermal cells, while in outer epidermal cells they seemed to be of less abundance.

Subcellular Analysis

Subcellular analysis was carried out on outer epidermal cells (Fig. 13). According to the spectra, four kinds of flavone derivatives were observed in the nucleus (*m/z* 463, 499, 625, and 661, *see* Fig. 13c). The cytoplasm had a more complex chemical composition. The four flavone derivatives that were already observed in the nucleus were also found in the cytoplasm, but with different abundances (Fig. 13b). Apart from the four species, other peaks of *m/z* 927, 1089, and 1251 were also observed, which had already been found in cellular analysis. The results clearly indicated that molecule distribution within a single cell was heterogeneous.

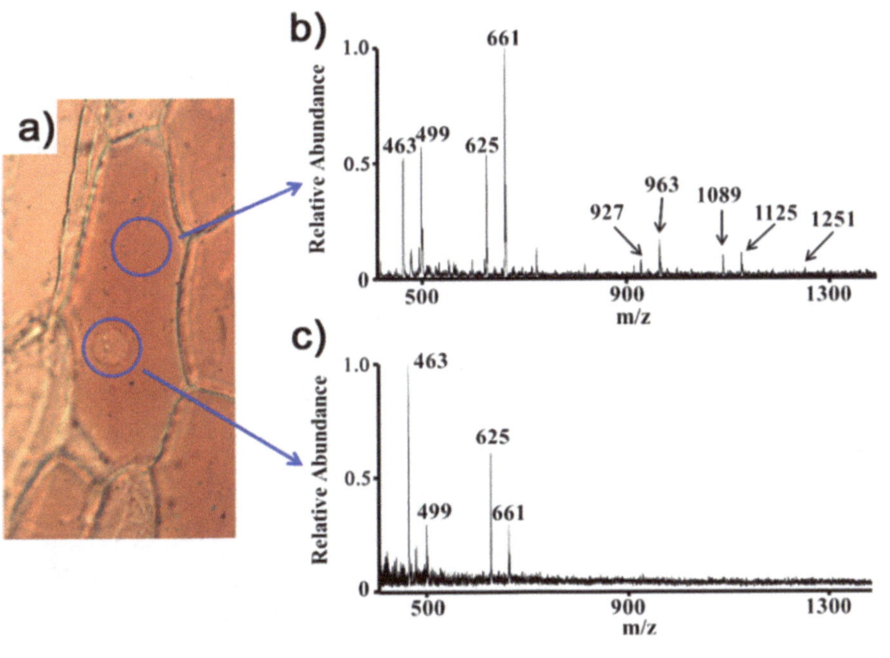

Fig. 13 Metabolite diversity of subcellular zones obtained from an outer epidermal cell. (**a**) Microscopic image of an outer epidermal cell; (**b**) Mass spectrum of the cytoplasm that was away from the nucleus; (**c**) Mass spectrum of the nucleus. The enrichment time was 30 s and the assistant solvent was methanol/water/chloroform = 3:1:1

4 Special Notice in Sampling Different Type of Cells

4.1 Sampling Cell into Emitter

The volume of a single cell varied from 1 pL to 10 nL. Therefore, the strategy for sampling different type of cell was different. *Allium cepa* epidermal cells are large cells (~ 1 nL) with only cell nucleus and vacuole. We can directly insert the emitter tip into the cell to sample hundreds of picoliter cell extract to the emitter tip. Moreover, the content was fluid with low viscosity which could be directly used for spray. Therefore, the content of *Allium cepa* cell could be directly used for Pulsed-DC-ESI-MS analysis without any further treatment (Fig. 14).

However, the mammalian cell had very viscous cytoplasm and small cell volume (~ 10 pL). The content of mammalian cell can not be directly used for spray. We had developed two strategies to deal with mammalian cells. One strategy is fully introduced and discussed in Subheading 3.4. Here, we introduce another strategy to sample whole cell into the emitter tip for Pulsed-DC-ESI analysis. We picked the target mammalian cell by single-cell manipulator with a sharp steel needle tip and placed the cell to tens of nanoliter spray solvent for cell lysis followed by Pulsed-DC-ESI-MS analysis (Fig. 15). As a result, the dilution degree of the sample was controlled by the spray solvent amount. For the consideration of current single-cell manipulating technique, we could at least use about 20 nL (Fig. 15) spray solvent for mammalian cell analysis (ensure the needle tip submerged by spray solvent). With the development of microfluidic techniques, we believe the amount of spray solvent could be reduced.

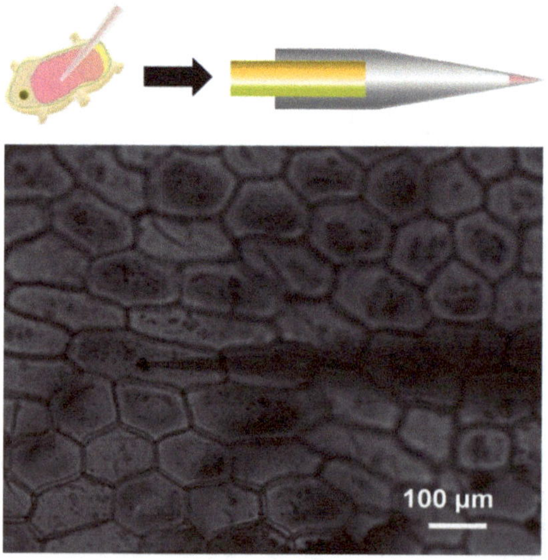

Fig. 14 Direct sampling single-cell contents into the emitter tip for Pulsed-DC-ESI-MS analysis. The emitter tip was controlled by the single-cell manipulator to insert into the cell and suck out the low viscous content. The contents were analyzed by Pulsed-DC-ESI-MS without any treatment

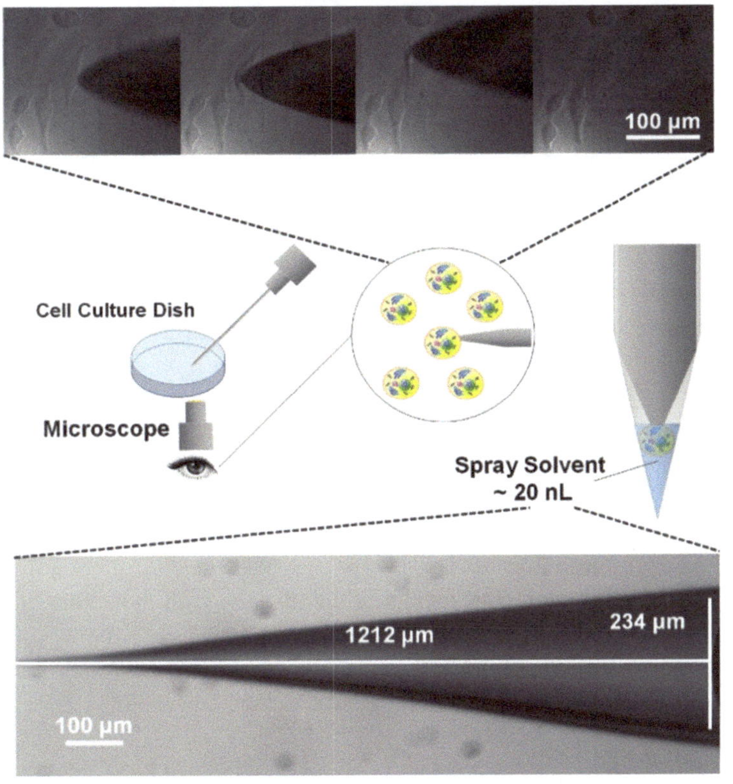

Fig. 15 Methods for sampling small volume single cell with viscous cytoplasm. In order to analyze this type of cells, cell lysis by spray solvent was essential. The target cell was picked by acupuncture needle tip from the culture dish by the single-cell manipulator. Then the needle tip was dipped into spray solvent preloaded in the emitter tip. Since the acupuncture tip could not be inserted to the very tip of the emitter, a certain amount of spray solvent was needed to submerge the tip. The lowest amount was about 20 nL

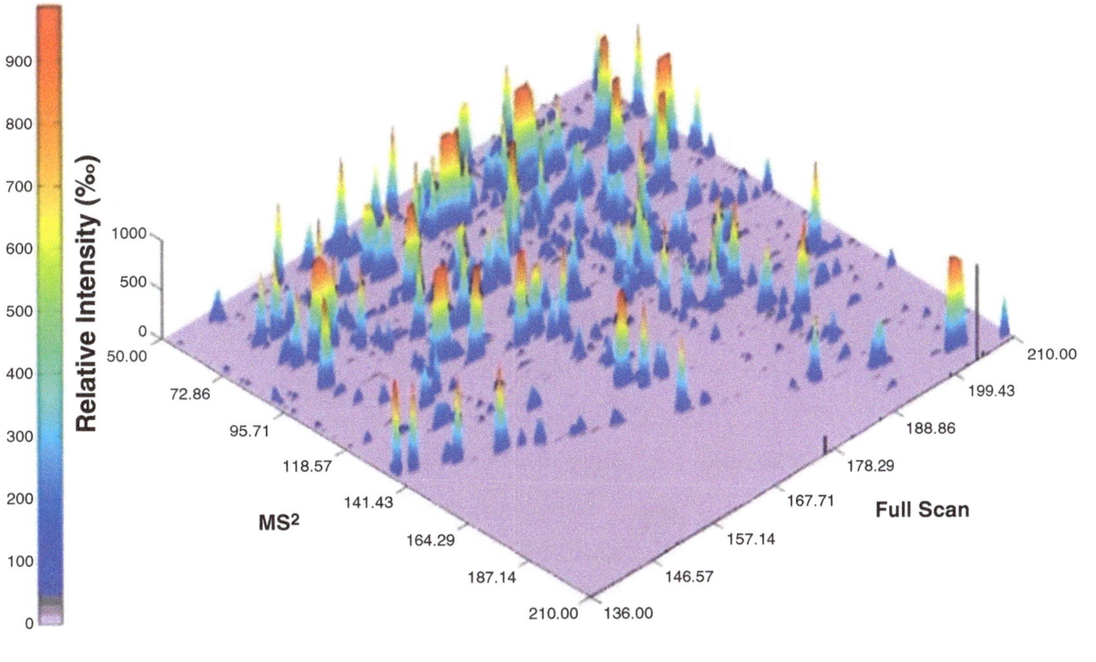

Fig. 16 The zoomed region (Full scan: 136 to 210, MS^2: 50 to 210) of *Allium cepa* cell 2-D mass spectrum

4.2 Single-Cell Metabolome Identification

The further identification of single-cell metabolome is dependent on the large scale of MS^2 data of the metabolite IDs (the 2-D mass spectrometry of single-cell metabolome, Fig. 2).

4.2.1 Database Search

For the small molecular weight metabolites, using MS^2 data for database match is a popular way for identification of metabolite IDs. This strategy is suitable for the identification of irregularly distributed peaks in Fig. 2. For example, when we perform Pulsed-DC-ESI analysis of single *Allium cepa* cell for metabolome identification, database search using massbank database (http://www.massbank.jp) was utilized for the identification of metabolites in the region: full scan mass number: 136 to 210; MS2 mass number: 50 to 210 (Fig. 16 and Table 2).

Table 2
The matching results of metabolites with molecular weight from 136 to 210 in Massbank database according to MS^2 data

m/z	Compound name	Massbank score
136.02	Adenine	0.7750
138.06	Trigonelline	0.9009
142.09	Histidinol	0.6170
145.05	Spermidine	0.6909
147.08	Lysine	0.9112
156.08	Histidine	0.9881
158.17	Arginine(Dehydration)	0.6102
159.07	Proline(Adimation)	0.7922
149.02	Glutamic acid	0.5313
162.11	Glucosamine(Dehydration)	0.6677
166.09	Phenylalanine	0.8007
171.02	Anserine(Metabolic Residues)	0.6912
175.12	Arginine	0.9599
178.05	Alliin	0.4460
180.09	Glucosamine	0.4934
185.03	Phosphocholine	0.5764
203.05	Arginine ethyl ester	0.8408
189.14	Arginine (Methylation)	0.6487
195.07	Caffeine	0.4334
198.10	Galactose	0.3938
205.10	Tryptophan	0.8762

4.2.2 Tracking a Certain Type of Metabolites and Identification of the Chemical Modification to Metabolites

In Fig. 2a, we could also intuitively find some peak groups along the MS^2 axis of 2-D mass spectrum. For instance, at the m/z of 185.04, 203.05, 347.10, and 365.11 on the MS^2 axis, peaks were more crowded than other lines. These regular distributed peaks always indicated a certain kind of metabolites with unique neutral loss or typical fragment value. For example, many saccharides in *Allium cepa* cell were in modified forms. By searching the neutral loss of 162 (basic saccharide unit [$C_6H_{12}O_6$–H_2O]), we could confirm 505 components containing saccharide structures from the 1034 components in a single cell (Fig. 6a). We searched all 505 components in a database considering the combination of saccharide oligomerization and different ion adducts (Table 3). We found 141 components, of which 17 components (12%) were

Table 3
The database considered all forms of unmodified saccharides with *m/z* below 2000

Saccharide Forms	*m/z*	Saccharide units number	Oligomerization
$[C_6H_{12}O_6 + Na]^+$	203.056	1	1
$[C_6H_{12}O_6 + K]^+$	219.030	1	1
$[C_{12}H_{22}O_{11} + Na]^+$	365.111	2	1
$[C_{12}H_{22}O_{11} + K]^+$	381.085	2	1
$[C_{12}H_{24}O_{12} + Na]^+$	383.122	1	2
$[C_{12}H_{24}O_{12} + K]^+$	399.096	1	2
$[C_{18}H_{32}O_{16} + Na]^+$	527.166	3	1
$[C_{18}H_{32}O_{16} + K]^+$	543.140	3	1
$[C_{18}H_{36}O_{18} + Na]^+$	563.188	1	3
$[C_{18}H_{36}O_{18} + K]^+$	579.162	1	3
$[C_{24}H_{42}O_{21} + Na]^+$	689.221	4	1
$[C_{24}H_{42}O_{21} + K]^+$	705.195	4	1
$[C_{24}H_{44}O_{22} + Na]^+$	707.232	2	2
$[C_{24}H_{44}O_{22} + K]^+$	723.206	2	2
$[C_{24}H_{48}O_{24} + Na]^+$	743.254	1	4
$[C_{24}H_{48}O_{24} + K]^+$	759.228	1	4
$[C_{30}H_{52}O_{26} + Na]^+$	851.276	5	1
$[C_{30}H_{52}O_{26} + K]^+$	867.250	5	1
$[C_{30}H_{60}O_{30} + Na]^+$	923.320	1	5
$[C_{30}H_{60}O_{30} + K]^+$	939.294	1	5
$[C_{36}H_{64}O_{32} + Na]^+$	1031.342	3	2

(continued)

Table 3 (continued)

Saccharide Forms	*m/z*	Saccharide units number	Oligomerization
$[C_{36}H_{64}O_{32} + K]^+$	1047.316	3	2
$[C_{36}H_{66}O_{33} + Na]^+$	1049.353	2	3
$[C_{36}H_{66}O_{33} + K]^+$	1065.327	2	3
$[C_{36}H_{72}O_{36} + Na]^+$	1103.386	1	6
$[C_{36}H_{72}O_{36} + K]^+$	1119.360	1	6
$[C_{42}H_{84}O_{42} + Na]^+$	1283.452	1	7
$[C_{42}H_{84}O_{42} + K]^+$	1299.426	1	7
$[C_{48}H_{84}O_{42} + Na]^+$	1355.452	4	2
$[C_{48}H_{84}O_{42} + K]^+$	1371.426	4	2
$[C_{48}H_{88}O_{44} + Na]^+$	1391.474	2	4
$[C_{48}H_{88}O_{44} + K]^+$	1407.448	2	4
$[C_{48}H_{96}O_{48} + Na]^+$	1463.518	1	8
$[C_{48}H_{96}O_{48} + K]^+$	1479.492	1	8
$[C_{54}H_{96}O_{48} + Na]^+$	1535.518	3	3
$[C_{54}H_{96}O_{48} + K]^+$	1551.492	3	3
$[C_{54}H_{108}O_{54} + Na]^+$	1643.584	1	9
$[C_{54}H_{108}O_{54} + K]^+$	1659.558	1	9
$[C_{60}H_{104}O_{52} + Na]^+$	1679.562	5	2
$[C_{60}H_{104}O_{52} + K]^+$	1695.536	5	2
$[C_{60}H_{110}O_{55} + Na]^+$	1733.595	2	5
$[C_{60}H_{110}O_{55} + K]^+$	1749.569	2	5
$[C_{60}H_{120}O_{60} + Na]^+$	1823.650	1	10
$[C_{60}H_{120}O_{60} + K]^+$	1839.624	1	10

oligosaccharides and 124 components (88%) were modified products of oligosaccharides in cell (Fig. 6b). Further identification figured out six modification types of saccharides in cell, including alkylation, amination, acylation, oxidation, small molecule adduct, and metal ion exchange (Fig. 17b), involving 28 chemical groups (Table 4). To our knowledge, this is the first time to examine and identify hundreds of metabolites and their modified products in a single cell, which could provide useful information for deep understanding of single-cell metabolome.

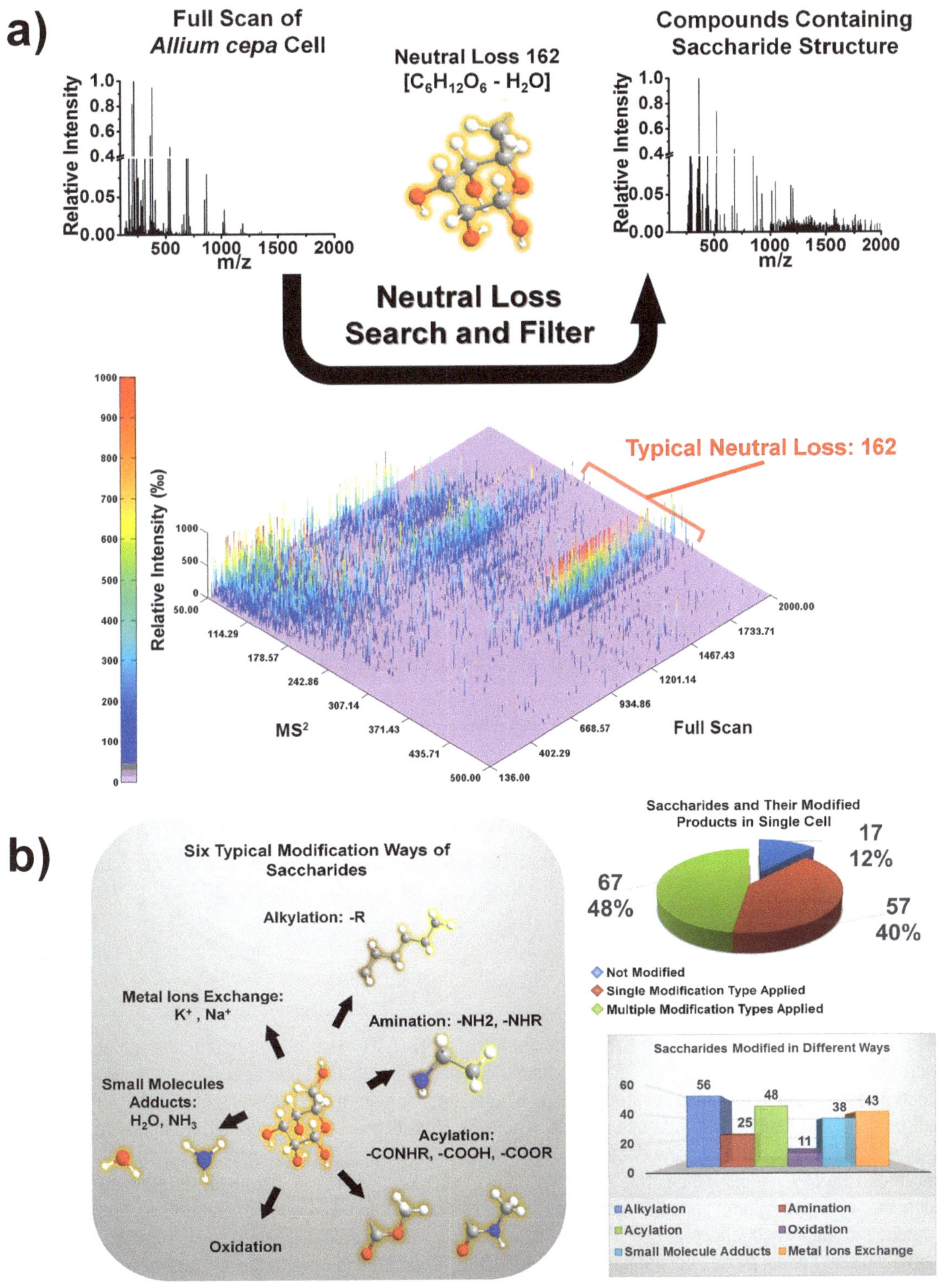

Fig. 17 Using neutral loss search to identify all saccharides and their modified products in a single cell. The typical neutral loss of saccharides and their modified products in MS/MS spectra was 162, which was the

Table 4
The 28 chemical groups participating in saccharide modification in single *Allium cepa* cell

Alkylation	Amination	Acylation	Oxidation	Small molecule adducts	Metal Ions Exchange
$-CH_2-$	-NH-	$-COCH_2-$	O	H_2O	Na
$-C_2H_2-$	$-NH_2$ (Substitution)	-CONH-		NH_3	K
$-C_2H_4-$	$-CH_2NH-$	-COO-			
$-C_3H_6$	$-C_2H_2NH-$	$-COOC_2H_4-$			
$-C_4H_6-$	$-C_3H_6NH-$	$-CONHC_2H_4-$			
$-C_4H_8-$	$-C_7H_{12}NH-$	$-COOC_3H_6-$			
$-C_6H_{10}-$	$-C_9H_{18}NH-$	$-COC_4H_8-$			
$-C_7H_{12}-$		$-COC_5H_{10}-$			

4.3 Safety Considerations

When experiments were performed, high D.C. voltage was applied to the copper wire. Actions for the insulation of both the operators and the MS from the high voltage source should be taken to avoid danger.

Acknowledgement

This research is supported by the 973 program (Grant 2013CB933800), the National Natural Science Foundation of China (Grants 21390411 and 21125525), and the Ministry of Science and Technology of China (Grants 2011YQ090005 and 2011YQ6008402).

Fig. 17 (continued) fragments of the monosaccharide unit ([$C_6H_{12}O_6$–H_2O]). After neutral loss search, we had screened out 505 components belonging to saccharides and their modified products. These components were then used for isotopic exclusion and accurate *m/z* matching in the database considered all forms of unmodified saccharides with *m/z* below 2000 (Table 3). Most of the components had a difference value (*D*-value) with the saccharides in the database. The *D*-value actually indicated the modified functional group. For example, the *D*-value of 14 indicated the methylation of saccharides (+CH_2). We finally determined 141 components, of which 17 components (12%) were oligosaccharides and 124 components (88%) were modified products of oligosaccharides in cell. We had figured out six modification types in cell, including alkylation, amination, acylation, oxidation, small molecule adducting, and metal ion exchange, which involved 28 chemical groups (Table 4)

References

1. Fiehn O (2002) Metabolomics--the link between genotypes and phenotypes. Plant Mol Biol 48:155–171
2. Holmes E, Wilson ID, Nicholson JK (2008) Metabolic phenotyping in health and disease. Cell 134:714–717
3. Dettmer K, Aronov PA, Hammock BD (2007) Mass spectrometry-based metabolomics. Mass Spectrom Rev 26:51–78
4. Fiehn O, Kopka J, Dormann P, Altmann T, Trethewey RN, Willmitzer L (2000) Metabolite profiling for plant functional genomics. Nat Biotechnol 18:1157–1161
5. Griffin JL, Shockcor JP (2004) Metabolic profiles of cancer cells. Nat Rev Cancer 4:551–561
6. Wilson ID, Nicholson JK, Castro-Perez J, Granger JH, Johnson KA, Smith BW, Plumb RS (2005) High resolution "ultra performance" liquid chromatography coupled to oa-TOF mass spectrometry as a tool for differential metabolic pathway profiling in functional genomic studies. J Proteome Res 4:591–598
7. Lisec J, Schauer N, Kopka J, Willmitzer L, Fernie AR (2006) Gas chromatography mass spectrometry-based metabolite profiling in plants. Nat Protoc 1:387–396
8. Beckonert O, Keun HC, Ebbels TMD, Bundy JG, Holmes E, Lindon JC, Nicholson JK (2007) Metabolic profiling, metabolomic and metabonomic procedures for NMR spectroscopy of urine, plasma, serum and tissue extracts. Nat Protoc 2:2692–2703
9. Doroghazi JR, Albright JC, Goering AW, Ju KS, Haines RR, Tchalukov KA, Labeda DP, Kelleher NL, Metcalf WW (2014) A roadmap for natural product discovery based on large-scale genomics and metabolomics. Nat Chem Biol 10:963–968
10. Rubakhin SS, Romanova EV, Nemes P, Sweedler JV (2011) Profiling signaling peptides in single mammalian cells using mass spectrometry. Nat Methods 8:S20–S29
11. Zenobi R (2013) Single-cell metabolomics: analytical and biological perspectives. Science 342:1201–120+
12. Wang F, Pan XH, Kalmbach K, Seth-Smith ML, Ye XY, Antumes DMF, Yin Y, Liu L, Keefe DL, Weissman SM (2013) Robust measurement of telomere length in single cells. P Natl Acad Sci USA 110:E1906–E1912
13. Ibanez AJ, Fagerer SR, Schmidt AM, Urban PL, Jefimovs K, Geiger P, Dechant R, Heinemann M, Zenobi R (2013) Mass spectrometry-based metabolomics of single yeast cells. P Natl Acad Sci USA 110:8790–8794
14. Rubakhin SS, Lanni EJ, Sweedler JV (2013) Progress toward single cell metabolomics. Curr Opin Biotech 24:95–104
15. Hummon AB, Sweedler JV, Corbin RW (2003) Discovering new neuropeptides using single-cell mass spectrometry. Trac-Trend Anal Chem 22:515–521
16. Lapainis T, Rubakhin SS, Sweedler JV (2009) Capillary electrophoresis with electrospray ionization mass spectrometric detection for single cell metabolomics. Anal Chem 81:5858–5864
17. Mizuno H, Tsuyama N, Harada T, Masujima T (2008) Live single-cell video-mass spectrometry for cellular and subcellular molecular detection and cell classification. J Mass Spectrom 43:1692–1700
18. Tejedor ML, Mizuno H, Tsuyama N, Harada T, Masujima T (2012) In situ molecular analysis of plant tissues by live single-cell mass spectrometry. Anal Chem 84:5221–5228
19. Stolee JA, Shrestha B, Mengistu G, Vertes A (2012) Observation of subcellular metabolite gradients in single cells by laser ablation electrospray ionization mass spectrometry. Angew Chem Int Ed 51:10386–10389
20. Wei Z, Xiong X, Guo C, Si X, Zhao Y, He M, Yang C, Xu W, Tang F, Fang X, Zhang S, Zhang X (2015) Pulsed direct current electrospray: enabling systematic analysis of small volume sample by boosting sample economy. Anal Chem 87(22):11242–11248
21. Wei ZW, Han S, Gong XY, Zhao YY, Yang CD, Zhang SC, Zhang XR (2013) Rapid removal of matrices from small-volume samples by step-voltage nanoelectrospray. Angew Chem Int Ed 52:11025–11028
22. Mandal MK, Saha S, Yoshimura K, Shida Y, Takeda S, Nonami H, Hiraoka K (2013) Biomolecular analysis and cancer diagnostics by negative mode probe electrospray ionization. Analyst 138:1682–1688
23. Yu Z, Chen LC, Suzuki H, Ariyada O, Erra-Balsells R, Nonami H, Hiraoka K (2009) Direct profiling of phytochemicals in tulip tissues and in vivo monitoring of the change of carbohydrate content in tulip bulbs by probe electrospray ionization mass spectrometry. J Am Soc Mass Spectrom 20:2304–2311

Chapter 5

Single-Probe Mass Spectrometry Analysis of Metabolites in Single Cells

Ning Pan, Wei Rao, and Zhibo Yang

Abstract

Mass spectrometry (MS) is an indispensable analytical technique for bioanalysis. Based on the measurement of mass/charge ratios (*m/z*) of ions, MS can be used for sensitive detection and accurate identification of species of interest. In traditional studies, MS is utilized to measure analytes in prepared solutions or gas-phase samples. Benefited from recent development of sampling and ionization approaches, MS has been extensively applied to the analysis of broad ranges of biological samples. We have developed a new device, the Single-probe, that can be used for in situ, real-time MS analysis of metabolites inside individual living cells. The Single-probe is a miniaturized multifunctional sampling and ionization device that is directly coupled to the mass spectrometer. With a sampling tip size smaller than 10 μm, we can insert the Single-probe tip into single cells to extract intracellular compounds, which are analyzed using MS in real-time. We have successfully used the Single-probe MS technique to detect a variety of endogenous and exogenous cellular metabolites in individual eukaryotic cells. Single cell mass spectrometry (SCMS) is a new scientific technology that has the potential to reshape approaches in biological and pharmaceutical bioanalytical research.

Key word Single-probe, Mass spectrometry, Metabolites, Single cell

1 Introduction

All living organisms are composed of a population of heterogeneous cells with different types and states. Understanding the living organisms at single cell level enables us to fundamentally investigate biological science, and potentially to revolutionize bioanalytical science and human medicine [1–3]. However, traditional bioanalysis techniques are generally based on the measurement of samples (e.g., cell lysate) prepared from mixed cells. Only averaged results can be obtained, eliminating the capability of acquiring crucial information from individual cells. Therefore, it is our goal to elucidate the behaviors and mechanisms in individual cells without drastically disrupting their environment or structure. Single cell analysis (SCA) is a rapidly emerging field in the metabolic

Bindesh Shrestha (ed.), *Single Cell Metabolism: Methods and Protocols*, Methods in Molecular Biology, vol. 2064,
https://doi.org/10.1007/978-1-4939-9831-9_5,

investigation of biological systems. Due to its high sensitivity and broad detection range, mass spectrometry (MS) has been extensively utilized for bioanalysis. Particularly, with recent developments in sampling and ionization techniques, MS has been used for the rapid and sensitive detection of molecular components at the cellular level [4–7]. A number of MS-based SCA methods have been reported, including secondary ion mass spectrometry (SIMS) [8–10], matrix-assisted laser desorption ionization (MALDI) [8, 11, 12], laser ablation electrospray ionization MS (LAESI) [13], laser desorption/ionization mass spectrometry (LDI-MS) [14, 15], capillary electrophoresis-electrospray ionization MS (CE-ESI-MS) [16–18], and Live Single-cell MS [19–23]. Although these established single-cell MS techniques have served well in a variety of studies, they contain inherent limitations that constrain their broader applications [10, 24–27]. These limitations include the requirement of high vacuum (e.g., SIMS and MALDI) and time-consuming sample pretreatment with large amount of matrix molecules, which dramatically alter the conditions of the biological samples, making it incompatible with live cells. Other requirements, such as extra fragments induced by laser ablation (e.g., SIMS, LAESI, and LDI-MS), reduced sensitivity due to the high concentration of salt in sample (e.g., CE-ESI-MS), and separated sample preparation (e.g., Live Single-cell MS), are also unfavored for live single-cell analysis. To overcome the drawbacks of existing MS-based techniques for SCA, we have developed a new device, the Single-probe, that can be coupled with MS for in situ live single cell detection in real-time. The Single-probe is a miniaturized, multifunctional sampling and ionization device. The name "Single-probe" is given due to its unity in design, multiple functions, and versatile applications, as it has also been successfully utilized for ambient MS imaging analysis [28].

2 Materials

2.1 Single-Probe Fabrication Protocol

The Single-probe is fabricated from several small pieces: (1) a laser-pulled dual-bore quartz needle, (2) a silica capillary, and (3) a nano-ESI emitter [29, 30] (Fig. 1a–c) (*See* **Note 1**).

1. Pull dual-bore quartz tubing (1.120″ × 0.005″ × 12″, MBT-005-020-2Q, Friedrich & dimmock, Inc., Millville, NJ) using a laser puller (Model P-2000, Sutter Instrument CO., Novato, CA). A dual-bore quartz needle with a sharp tip (<10 μm) is produced (Fig. 1c).
2. The Single-probe is fabricated by embedding a fused silica capillary (ID: 40 μm, OD: 110 μm, Polymicro Technologies, Phoenix, AZ) as the solvent-providing capillary and a nano-ESI emitter (pulled from the same type of fused silica capillary) into

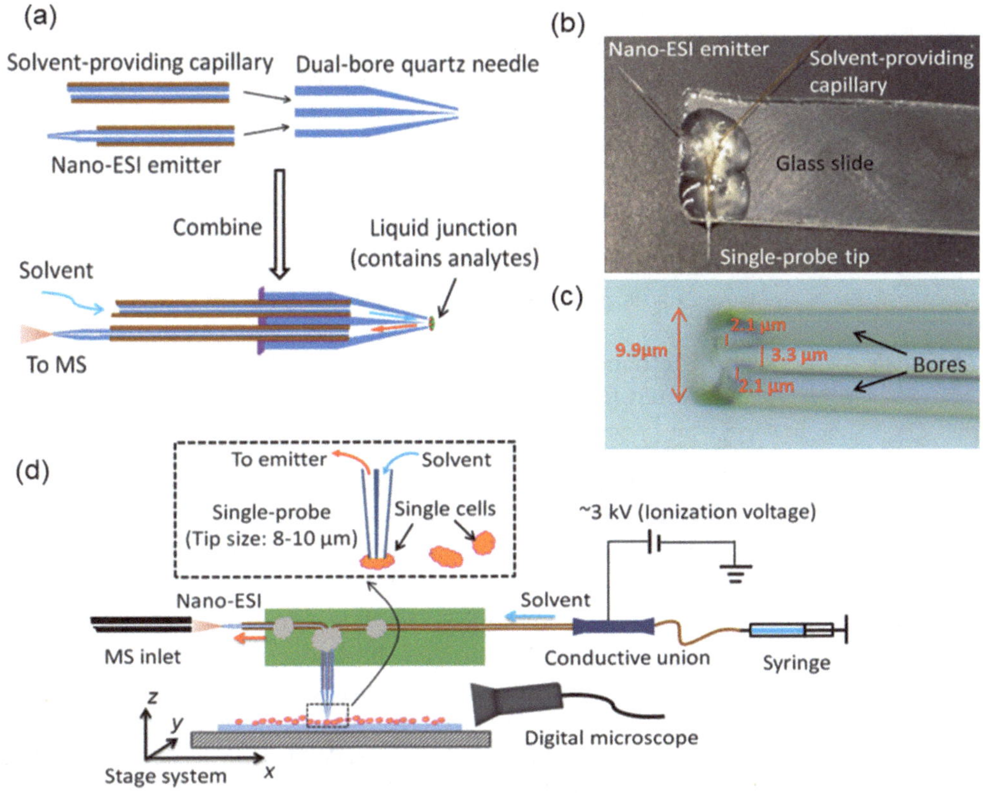

Fig. 1 The Single-probe MS system from fabrication to utilization. (**a**) Fabrication steps of the Single-probe; (**b**) photograph of a working Single-probe; (**c**) 40× magnification of the Single-probe tip with measurements obtained from the calibrated digital microscope; (**d**) setup schematic of a single-cell MS analysis

the laser pulled dual-bore quartz needle. UV-curing epoxy is used as the adhesive (item No. 006.030, Prime Dental, Chicago, IL) (Fig. 1 a–c). To cure the UV epoxy, irradiate the glued positions with a LED UV lamp (e.g., LY-C240, Foshan Liang Ya Dental Equipment Co., Ltd., Guangdong, China) for ~30 s (*See* **Note 2**).

3. To conveniently use the device, glue the Single-probe onto a microscope glass slide using regular epoxy (e.g., 5 min Epoxy, Part No. 20945, ITW Devcon, Inc., Danvers, MA) (Fig. 1b).
4. An inline MicroFilter (M-520, IDEX Health & Science LLC, Oak Harbor, WA) is attached to the syringe through a fused silica capillary (ID: 50 μm, OD: 360 μm, Polymicro Technologies, Phoenix, AZ) to filtrate the solvent before being delivered to the Single-probe.
5. A Conductive MicroTight Union (M-539, IDEX Health & Science, LLC, Oak Harbor, WA) is connected between the fused silica capillary (ID: 50 μm, OD: 360 μm) and the solvent-providing capillary (ID: 40 μm, OD: 110 μm) on the

Single-probe; PEEK sleeve tubings with different sizes (F-181 and F-380, IDEX, IDEX Health & Science, LLC, Oak Harbor, WA) are needed to make sealed connection. To conduct the experiment, the ionization voltage is applied on the union instead of the nano-ESI emitter, which is very short (<5 mm) and non-conductive.

6. Syringe (100 mL, 250 mL, 500 mL, Hamilton, Inc., Reno, NV).
7. Nexus 3000 Syringe Pump (Chemyx, Inc., Stafford, TX).

2.2 Chemicals

1. Methanol (Fluka Analytical, LC-MS Ultra CHROMASOLV, tested for UHPLC-MS, Sigma-Aldrich, St. Louis, MO).
2. H_2O Water (Lot RNBD4563, Sigma-Aldrich, St. Louis, MO).
3. Acetonitrile (Fluka Analytical, LC-MS Ultra CHROMASOLV, tested for UHPLC-MS, Sigma-Aldrich, St. Louis, MO).
4. Acetic acid (Glacial, >99.85%, Sigma-Aldrich, St. Louis, MO).

2.3 Materials for Cell Culture

2.3.1 Complete Cell Culture Medium

1. Dulbecco's Modified Eagle's Medium (DMEM) (Cellgro, Manasas, VA) with 10% heat-inactivated fetal bovine serum (FBS) (Gibco/Life Technologies, Long Island, NY), Penicillin/Streptomycin (Cellgro, Manasas, VA), and 10 mM HEPES (pH 7.4) (Cellgro, Manasas, VA). FBS is heat inactivated by incubation at 55 °C for 15 min.
2. 1× Phosphate Buffered Saline (PBS) (Cellgro, Manassas, VA).
3. TrypLE™ Express (Gibco/Life Technologies, Long Island, NY).
4. 12-well plates and T25 flasks (Sterile individually wrapped, Corning Costar, Sigma-Aldrich, St. Louis, MO).
5. 18 mm round cover slides (Autoclaved, VWR, Radnor, PA).

2.3.2 Cell (HeLa) Culture

HeLa cells are selected as the model system for the study. All cell culture procedures are performed under sterile conditions in a biological safety cabinet. Cells are cultured in complete DMEM medium as follows:

1. All reagents are warmed to 37 °C prior to use.
2. HeLa cells initially grow to a confluence of 70–80%.
3. The growth media is aspirated and cells are briefly rinsed with 5 mL of 1× PBS.
4. After removal of 1× PBS, cells are treated with 2.5 mL of TrypLE Express and incubated at 37 °C for 10 min (Heracell 240, Thermo Fisher Scientific, Inc., Waltham, MA).

5. The activity of TrypLE express is stopped with 7.5 mL of complete media. The cells are resuspended to make a uniform suspension.
6. In a 12-well plate, single cover-slides are placed and 1.8 mL of DMEM media is added; additional 0.2 mL of cell suspension is added to the same well. The cells are mixed by gentle rocking of the plate and incubated at 37 °C with 5% CO_2.
7. To culture cells, 1 mL of cell suspension is added to 9 mL of complete DMEM and transferred to a T25 flask, and cells are incubated at 37 °C with 5% CO_2. The passage number of the cells was noted each time during culturing of cells.

3 Methods

3.1 Single-Probe MS Setup

The Single-probe SCMS experimental system includes a digital stereo microscope (Supereyes T004 Digital Microscope, Shenzhen D&F Co., Ltd., Shenzhen, China), a USB digital microscope ("Scorpio" 500× Zoom, Chinavasion Wholesale Ltd., Shenzhen, China), a motorized X,Y,Z–translation stage system (CONEX-MFACC, Newport Corp., Irvine, CA), a manual X,Y, Z–translation stage (Compact Dovetail XYZ Linear Stage, Newport Corp., Irvine, CA), a flexible holder (MXB-3 h, Siskiyou Corp., Grants Pass, OR), and a Thermo LTQ Orbitrap XL mass spectrometer (Thermo Fisher Scientific, Inc., Waltham, MA) (Fig. 2) (*See* **Note 3 and 4**).

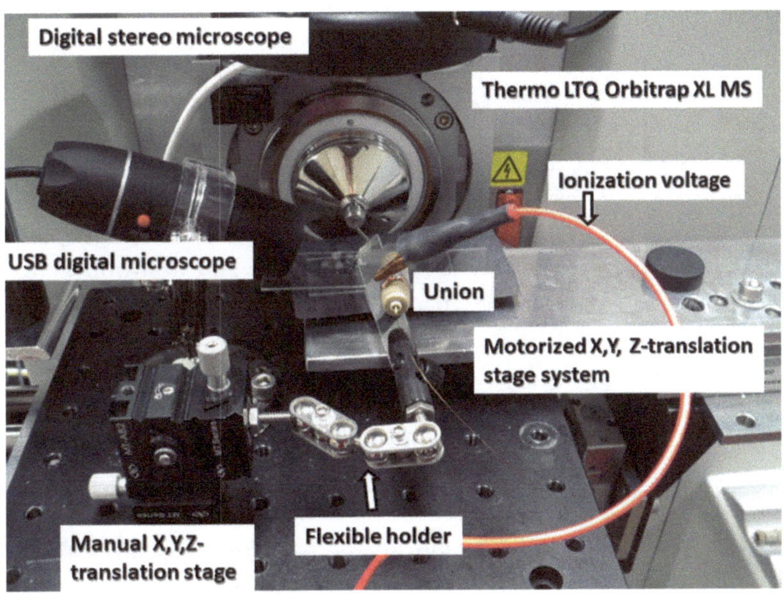

Fig. 2 Single-probe SCMS setup

1. Single-probe is attached to a microscope slide, which is held by the Siskiyou flexible holder (attached to the manual X,Y,Z-translation stage). Adjust the position of the Single-probe such that the nano-ESI emitter is at an optimized position toward the mass spectrometer inlet.
2. Fill the Hamilton syringe (e.g., 250 μL) with sampling solvent (MeOH/H_2O or acetonitrile with 1% acetic acid). Attach the syringe to an inline MicroFilter through a fused silica capillary (ID: 50 μm, OD: 360 μm), which is connected with a Conductive Microunion (IDEX Health & Science LLC, M-539). The ionization voltage (~3 kV) is applied on the Microunion, and transmitted through the solvent (inside the solvent-providing capillary and the dual-bore quartz needle) to the nano-ESI emitter.
3. Focus the digital stereo microscope on the tip of the Single-probe; this microscope will be used to monitor the penetration of the entire single cell analysis.
4. The USB digital microscope on the side is used to monitor the position of the nano-ESI emitter on the Single-probe.
5. Attach the cell sample to the motorized X,Y,Z-stage system, and the LabView software package [31] is used to control the position of the cells.

3.2 Experiment Protocols

1. Once the Single-probe is setup, turn on the syringe pump to deliver the sampling solvent. The ionization voltage (~3 kV) is applied on the Conductive MicroTight Union, and solvent ion signals can be immediately observed, indicating that the Single-probe is working properly. If there is no signal observed, the Single-probe is not functioning properly. Possible reasons include that clogging in the Single-probe and or out-of-position of the nano-ESI emitter. Solvent flow rate is critical for the SCMS measurement. In general, the flow rate is around 25 nL/min, depending on the tip sizes of the dual-bore quartz needle and the nano-ESI emitter. The optimized flow rate is achieved when there are no droplets dripping from dual-bore quartz needle tip and no bubbles inside the nano-ESI emitter.
2. Prior to the experiment, the cells are rinsed with PBS (phosphate buffered saline) or cell culture medium to remove the extracellular species that can affect the detection of species of interest (e.g., drugs used in the treatment). When the Single-probe works well, place the cell-containing plate on the X,Y,Z-stage underneath the Single-probe tip, guided by the digital stereo microscopes.
3. Upon the selection of an individual cell of interest, the Z-translation stage is lifted precisely to insert the Single-probe tip into the cell. The motorized stages used in the

setup allows for precise control (in 0.1 μm increments). Cell penetration by the Single-probe tip can be monitored using the stereo microscope.

4. A time delay of 1–2 s is usually observed between the moment of insertion and detection of the MS signal. A dramatic change of the MS signals can be observed upon the penetration of the cell membrane. Single-probe sampling continues till no cellular signals are observed. The Single-probe is then lifted, and the constant flow of the sampling solvent flushes the channels inside of the probe to eliminate carryover effects. Each cell experiment requires ~3 min.
5. Sometimes, despite the flushing of solvent, probe can be clogged. To solve the clogging problem, a miniature heating coil (ID = 1 mm, Cr-Ni filament, gauge 34) is fabricated to heat up the clogged positions (e.g., dual-bore quartz needle tip coil or nano-ESI emitter) for several seconds, and then flow the solvent through the probe to remove the debris. If clogging cannot be resolved, the probe needs to be replaced.

3.3 Results

The Single-probe MS technique has been successfully used to analyze individual HeLa cells [32]. Cellular metabolites in the individual cells were detected (Fig. 3). In the positive ion mode, we have identified 18 different lipid species, including phosphatidylcholines (PC), sphingomyleins (SM), diglycerides (DG), and triglycerides (TG) [33]. In the negative ion mode, we have detected adenosine phosphates (AMP, ADP, and ATP). In the experiments of cells treated with the anticancer drugs doxorubicin (Adriamycin) [34, 35], paclitaxel (Taxol) [36, 37], or OSW-1 [38–40], we have observed the corresponding drugs and their metabolites in individual HeLa cells. To prepare for the cell samples, HeLa cells were cultured and dosed with anticancer compounds for 4 h at different concentrations (i.e., 10 nM, 100 nM, 1 μM, 10 μM) before SCMS measurement. Untreated HeLa cells served as controls.

4 Notes

1. For Single-probe fabrication, glue two capillaries inside of the dual-bore quartz tubing, make sure the glue does not seal the capillary inlet. Ware UV protection safety glass when using the UV lamp.
2. Nano-ESI emitter tip size should be similar to the Single-probe tip size.
3. Before running a single cell MS experiment, test the Single-probe using the sampling solvent to check if the probe functions as expected.

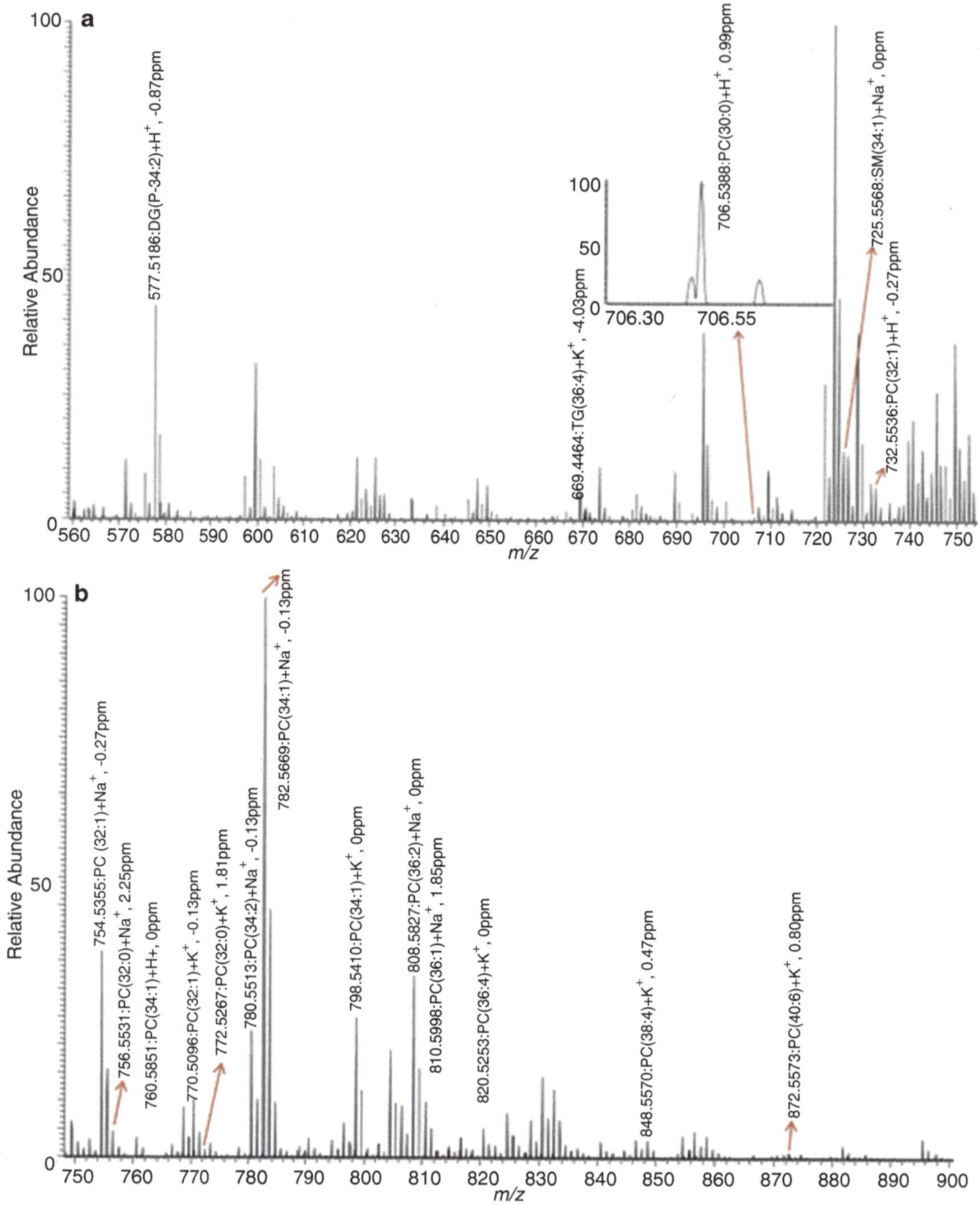

Fig. 3 Mass spectra of single HeLa cells from the Single-probe MS analysis (**a**) Mass range: *m/z* 560-*m/z* 750, (**b**) Mass range: *m/z* 750-*m/z* 900. Identified compounds are labeled with measured mass, compound name, and mass deviation (*Note*: Figs. 1 and 3 are the re-prints from our publication, N. Pan, W. Rao, N.R. Kothapalli, R. Liu, A.W.G. Burgett, and Z. Yang, *Analytical Chemistry* 2014, *86*, 9376–9380)

4. In instances where the Single-probe becomes obstructed, we used a homebuilt mini-heating coil to heat up the probe tip and nano-ESI emitter to clean them up. This device has four major components: a glass slide, two copper wires to mount the resistance heating wire on the glass slide, a coiled resistance heating wire (Nickel-Chromium alloy wire, gauge 34) attached to the copper wires, and a variable AC transformer to provide heating power. To operate, ~8 V potential is applied to heat the coil (glowing red). The clogged parts (usually the Single-probe tip and nano-ESI emitter) is inserted into the coiled wire and kept steady for several seconds. After unclogging, the debris remaining in the Single-probe is flushed with solvent and the probe is ready to be reused.

Acknowledgments

The authors would like to acknowledge Dr. Shaorong Liu at the University of Oklahoma for his generous support of our efforts, including the training and use of his laser micropipette puller. We would also like to thank Dr. Julia Laskin at the Pacific Northwest National Laboratory (PNNL) for her consultation and sharing the translation stage control program. This research was supported by grants from the Research Council of the University of Oklahoma Norman Campus, the American Society for Mass Spectrometry Research Award (sponsored by Waters Corporation), and Oklahoma Center for the Advancement of Science and Technology (grant HR 14-152).

References

1. Andersson H, van den Berg A (2004) Microtechnologies and nanotechnologies for single-cell analysis. Curr Opin Biotechnol 15 (1):44–49
2. Wang DJ, Bodovitz S (2010) Single cell analysis: the new frontier in 'omics'. Trends Biotechnol 28(6):281–290
3. Rubakhin SS, Romanova EV, Nemes P, Sweedler JV (2011) Profiling metabolites and peptides in single cells. Nat Methods 8(4): S20–S29
4. Dominguez MH, Chattopadhyay PK, Ma S, Lamoreaux L, McDavid A, Finak G, Gottardo R, Koup RA, Roederer M (2013) Highly multiplexed quantitation of gene expression on single cells. J Immunol Methods 391(1–2):133–145
5. Powell AA, Talasaz AH, Zhang HY, Coram MA, Reddy A, Deng G, Telli ML, Advani RH, Carlson RW, Mollick JA, Sheth S, Kurian AW, Ford JM, Stockdale FE, Quake SR, Pease RF, Mindrinos MN, Bhanot G, Dairkee SH, Davis RW, Jeffrey SS (2012) Single cell profiling of circulating tumor cells: transcriptional heterogeneity and diversity from breast cancer cell lines. PLoS One 7(5)
6. Citri A, Pang ZPP, Sudhof TC, Wernig M, Malenka RC (2012) Comprehensive qPCR profiling of gene expression in single neuronal cells. Nat Protoc 7(1):118–127
7. Narsinh KH, Sun N, Sanchez-Freire V, Lee AS, Almeida P, Hu SJ, Jan T, Wilson KD, Leong D, Rosenberg J, Yao M, Robbins RC, Wu JC (2011) Single cell transcriptional profiling reveals heterogeneity of human induced pluripotent stem cells. J Clin Investig 121 (3):1217–1221
8. Pacholski ML, Winograd N (1999) Imaging with mass spectrometry. Chem Rev 99 (10):2977–3006

9. Chughtai K, Heeren RMA (2010) Mass spectrometric imaging for biomedical tissue analysis. Chem Rev 110(5):3237–3277
10. Lanni EJ, Rubakhin SS, Sweedler JV (2012) Mass spectrometry imaging and profiling of single cells. J Proteome 75(16):5036–5051
11. Berry KAZ, Hankin JA, Barkley RM, Spraggins JM, Caprioli RM, Murphy RC (2011) MALDI imaging of lipid biochemistry in tissues by mass spectrometry. Chem Rev 111(10):6491–6512
12. Amantonico A, Oh JY, Sobek J, Heinemann M, Zenobi R (2008) Mass spectrometric method for analyzing metabolites in yeast with single cell sensitivity. Angewandte Chemie-International Edition 47 (29):5382–5385
13. Nemes P, Vertes A (2007) Laser ablation electrospray ionization for atmospheric pressure, in vivo, and imaging mass spectrometry. Anal Chem 79(21):8098–8106
14. Greving MP, Patti GJ, Siuzdak G (2011) Nanostructure-initiator mass spectrometry metabolite analysis and imaging. Anal Chem 83(1):2–7
15. Urban PL, Schmid T, Amantonico A, Zenobi R (2011) Multidimensional analysis of single algal cells by integrating microspectroscopy with mass spectrometry. Anal Chem 83 (5):1843–1849
16. Mellors JS, Jorabchi K, Smith LM, Ramsey JM (2010) Integrated microfluidic device for automated single cell analysis using electrophoretic separation and electrospray ionization mass spectrometry. Anal Chem 82(3):967–973
17. Nemes P, Rubakhin SS, Aerts JT, Sweedler JV (2013) Qualitative and quantitative metabolomic investigation of single neurons by capillary electrophoresis electrospray ionization mass spectrometry. Nat Protoc 8(4):783–799
18. Nemes P, Knolhoff AM, Rubakhin SS, Sweedler JV (2011) Metabolic differentiation of neuronal phenotypes by single-cell capillary electrophoresis-electrospray ionization-mass spectrometry. Anal Chem 83(17):6810–6817
19. Fukano Y, Tsuyama N, Mizuno H, Date S, Takano M, Masujima T (2012) Drug metabolite heterogeneity in cultured single cells profiled by pico-trapping direct mass spectrometry. Nanomedicine-Uk 7(9):1365–1374
20. Masujima T (2009) Live single-cell mass spectrometry. Anal Sci 25(8):953–960
21. Mizuno H, Tsuyama N, Harada T, Masujima T (2008) Live single-cell video-mass spectrometry for cellular and subcellular molecular detection and cell classification. J Mass Spectrom 43(12):1692–1700
22. Tsuyama N, Mizuno H, Tokunaga E, Masujima T (2008) Live single-cell molecular analysis by video-mass spectrometry. Anal Sci 24 (5):559–561
23. Tejedor ML, Mizuno H, Tsuyama N, Harada T, Masujima T (2012) In situ molecular analysis of plant tissues by live single-cell mass spectrometry. Anal Chem 84 (12):5221–5228
24. Miura D, Fujimura Y, Wariishi H (2012) In situ metabolomic mass spectrometry imaging: recent advances and difficulties. J Proteome 75(16):5052–5060
25. Passarelli MK, Ewing AG (2013) Single-cell imaging mass spectrometry. Curr Opin Chem Biol 17(5):854–859
26. Boggio KJ, Obasuyi E, Sugino K, Nelson SB, Agar NYR, Agar JN (2011) Recent advances in single-cell MALDI mass spectrometry imaging and potential clinical impact. Expert Rev Proteomics 8(5):591–604
27. Tsuyama N, Mizuno H, Masujima T (2011) Mass spectrometry for cellular and tissue analyses in a very small region. Anal Sci 27 (2):163–170
28. Rao W, Pan N, Yang Z (2015) High resolution tissue imaging using the single-probe mass spectrometry under ambient conditions. J Am Soc Mass Spectrom 26(6):986–993
29. Wilm M, Mann M (1996) Analytical properties of the nanoelectrospray ion source. Anal Chem 68(1):1–8
30. Wilm M, Neubauer G, Mann M (1996) Parent ion scans of unseparated peptide mixtures. Anal Chem 68(3):527–533
31. Lanekoff I, Heath BS, Liyu A, Thomas M, Carson JP, Laskin J (2012) Automated platform for high-resolution tissue imaging using nanospray desorption electrospray ionization mass spectrometry. Anal Chem 84 (19):8351–8356
32. Pan N, Rao W, Kothapalli NR, Liu R, Burgett AWG, Yang Z (2014) Anal Chem 86:9376–9380
33. Schober Y, Guenther S, Spengler B, Rompp A (2012) Single cell matrix-assisted laser desorption/ionization mass spectrometry imaging. Anal Chem 84(15):6293–6297
34. Lal S, Mahajan A, Chen WN, Chowbay B (2010) Pharmacogenetics of target genes across doxorubicin disposition pathway: a review. Curr Drug Metab 11(1):115–128

35. Chlebowski RT (1979) Adriamycin (doxorubicin) cardiotoxicity - review. West J Med 131 (5):364–368
36. Spencer CM, Faulds D (1994) Paclitaxel - a review of its pharmacodynamic and pharmacokinetic properties and therapeutic potential in the treatment of cancer. Drugs 48(5):794–847
37. Marupudi NI, Han JE, Li KW, Renard VM, Tyler BM, Brem H (2007) Paclitaxel: a review of adverse toxicities and novel delivery strategies. Expert Opin Drug Saf 6(5):609–621
38. Burgett AWG, Poulsen TB, Wangkanont K, Anderson DR, Kikuchi C, Shimada K, Okubo S, Fortner KC, Mimaki Y, Kuroda M, Murphy JP, Schwalb DJ, Petrella EC, Cornella-Taracido I, Schirle M, Tallarico JA, Shair MD (2011) Natural products reveal cancer cell dependence on oxysterol-binding proteins. Nat Chem Biol 7(9):639–647
39. Mimaki Y, Kuroda M, Takaashi Y, Sashida Y (1997) Concinnasteoside a, a new bisdesmosidic cholestane glycoside from the stems of Dracaena concinna. J Nat Prod 60(11):1203–1206
40. Zhou Y, Garcia-Prieto C, Carney DA, Xu RH, Pelicano H, Kang Y, Yu WS, Lou CG, Kondo S, Liu JS, Harris DM, Estrov Z, Keating MJ, Jin ZD, Huang P (2005) OSW-1: a natural compound with potent anticancer activity and a novel mechanism of action. J Natl Cancer Inst 97(23):1781–1785

Chapter 6

Applications of MicroArrays for Mass Spectrometry (MAMS) in Single-Cell Metabolomics

Alfredo J. Ibáñez and Ales Svatos

Abstract

The metabolic network is the endpoint in the flow of information that begins with the "gene" and ends with "phenotype" (observable function) of the cell. Previously, due to the variety of metabolites analyzed inside cells, the metabolomic measurements were performed with samples including multiple cells. Unfortunately, this sampling process may mask important metabolic phenomena, such as cell-to-cell heterogeneity. For these studies, we must use analytical techniques that can robustly deliver reproducible results with single-cell sensitivity. In this chapter, we summarize laser-based methods for single-cell analysis and a novel approach of MicroArrays for Mass Spectrometry (or MAMS) is described in full detail. This particular type of microarrays was tailored for the study of cells grown in liquid medium using multiple-analytical readouts, such as optical and laser desorption/ionization (LDI) or MALDI mass spectrometry.

Key words Single cell, Cell populations, Heterogeneity, Metabolites, Mass spectrometry, Laser, MALDI, Microarray, MAMS

1 Introduction

1.1 Features of Cell Population Metabolomics

There is currently a rediscovery of the metabolome importance in biological studies due to the advent of better analytical technologies. Metabolites carry out multiple functions in a cell; they are energy/fuel sources, regulators of enzyme activity, defense compounds, etc. For example, alterations in the central metabolism are the first reaction mechanism that eukaryotic cells have developed to counteract deleterious consequences that affect the cell's response to oxidative stress [1]. Furthermore, the metabolic network is the endpoint of the flow of biological information that starts with the "Gene" and ends with the "observable phenotype" (function) of the cell. Metabolic networks can accurately reflect the biochemical phenotype of a cell [2].

Interestingly, today (the era of antibiotic-resistance bacteria and metabolic diseases), metabolomics workflows are based on homogenizing a cell population prior to the elucidation of its

Bindesh Shrestha (ed.), *Single Cell Metabolism: Methods and Protocols*, Methods in Molecular Biology, vol. 2064, https://doi.org/10.1007/978-1-4939-9831-9_6, © Springer Science+Business Media, LLC, part of Springer Nature 2020

biochemical status and its phenotypic traits. This analytical workflow has not changed since the late 1950s, when multiple studies demonstrated that the processes within the cells are rarely identical even between the different individuals within an isogenic cell population [3].

Cell-to-cell (phenotypic) heterogeneity is a naturally occurring phenomenon that manifests itself in all organisms because it ensures enhanced adaptability to fast changes and/or perturbations in the growth environment. Several causes can induce cell-to-cell heterogeneity. Such variation has medical relevance, as for example in the case of drug response and persistence. It has been observed that the average response from a cell population toward a stimulus may not occur uniformly in each cell [4]. This is the main reason for persistence to a drug, which means that a small group of cells can better endure the addition of the drug to its growth medium (i.e., because they show a better phenotypic adaptation toward the drug than the rest of the cell population) [5]. Once the drug is removed, the selection pressure disappears and these surviving cells would give rise to a new (heterogeneous)—but still isogenic—population [5]. Thus, cell-to-cell heterogeneity must be taken into account when developing a mathematical model of the adaptation and survival mechanisms of a cell population toward a drug. It is for this reason that new analytical methods for identifying such cell-to-cell phenotypic variations are greatly required [5–8].

1.2 Laser-Based Methods for Sample Preparation-Analysis

The classical analytical workflow relies on sample dissection, solvent or thermal. The classical analytical workflow relies on sample dissection, solvent or thermal extraction and subsequent analysis using chromatographic methods hyphenated to a detector with particularly needed selectivity. Minimal sample preparation will make the analysis suitable to more laboratories and will allow people with minimal training in analytical chemistry to perform the analysis they need to perform without high knowledge barrier. Fortunately, a vast arsenal of different ionization techniques has been developed. Recent reviews [9–11] summarize established techniques for most of the possible applications to date, an excellent opportunity for beginners to get into the field. Given the amount of emerging, new techniques, an effort seems to be made in developing ionization techniques working with a minimal amount of sample preparation, and at ambient/atmospheric pressure. These techniques are readily applied in the life sciences [12, 13].

Another feature becoming very prominent in life sciences in general is mass spectrometry imaging (MSI) in high lateral or spatial resolution. Although light microscopy can already do this for a long time, MSI combines spatial information with full-fletched MS data [14, 15] suitable for compound characterization and discovery. However, light microscopy already can resolve subcellular to the point of finding a single molecule within a living cell in vivo. The only MS technique with a lateral resolution below the μm regime available is secondary ion mass spectrometry (SIMS) [16]. Using an ion beam to create secondary ions from the sample, it is not considered a soft ionization technique. Analytes tend to fragment upon ionization and the utilization of SIMS is intrinsically linked with extensive sample preparation. SIMS has been used on biological samples for imaging [14]. MSI of intact biomolecules however struggled to reach the bacterial cell level. Recent advances report single cell resolution with matrix-assisted ionization techniques, involving extensive sample preparation prior to analysis [17]. In early 2015 single-cell imaging was done within a tissue [18] utilizing ambient ionization, requiring considerable less sample preparation.

An ionization technique very prominent in the field of large molecule imaging is the matrix-assisted laser desorption/ionization (MALDI) [10, 15, 18]. It requires the samples to be preprocessed extensively by dissolution in and co-crystallization together with a matrix as well as it typically requires a vacuum to work [19–21]. Desorption and ionization of sample:matrix crystals is facilitated by an ultraviolet (UV) laser and more recently with infrared (IR) lasers. The matrix molecules absorb most of the energy deposited by the laser and transfer it to the sample analytes more gently [15, 22]. This has allowed MALDI to ionize and keep intact large bio-molecules, for example, proteins MALDI has been adapted to work under atmospheric pressure [23, 24]. MSI with MALDI is at the stage where instrumentation is commercially available with a lateral resolution of 10 μm (FLEX series, Bruker, USA). Restrains of the method are the need for reliable matrix deposition and ion yield of MALDI [10, 22]. For imaging plant cells, which are as large as 50 μm, this lateral resolution is sufficient. Early work [25] has demonstrated the utility of LDI-MSI for glandular pigment cells of *Hypppericum perforatum* or petal/sepal of *Arabidopsis thaliana* without using matrix. The plant compounds of polyketide origin are absorbing UV-laser energy and serve to desorb/ionize themselves. This method is very suitable, as the sample is not pre-treated with matrix and the lateral resolution is not compromised by matrix crystals, which could be larger than the studied cells. However, working at high vacuum, biological samples are losing water and the shape changes are no evitable. Working under ambient conditions is now possible using commercial AP-SMALDI10 (TransMIT, Giessen, Germany) which combined high lateral resolution with high mass accuracy and resolution. The lateral resolution is suitable for imaging single cells [26].

Laser ablation combined with inductive plasma ionization (ICP-MS) is used for imaging distribution of metals *in planta* [27] or to localize proteins labeled with antibodies containing a metal-reporter ion [28]. This matrix-free method shows extreme sensitivity and as desorbed tissue debris undergoes post-ionization in a plasma torch, the technique is also quantitative.

Another technique, matrix-assisted laser desorption electrospray ionization (MALDESI), separates the sample desorption/ablation step from ionization [29]. Being a MALDI—Electrospray ionization (ESI) hybrid technique, MALDESI accomplishes a higher ion yield because of the post-ionization with ESI, but still requires the deposition of an external matrix. While imaging lipid distributions across rat brains, the lateral resolution of 45 μm has been reported, applying oversampling [30].

ESI was originally designed to ionize long polymer chains and has subsequently evolved to a commonly used ion source in mass spectroscopy. ESI has become very popular [9], for example in combination with liquid chromatography and been used for MSI as well, especially in the form of desorption electrospray ionization (DESI, [31]) and the closely related nano-DESI [32] have been shown to achieve 50 and 20 μm lateral resolution, respectively [32, 33]. Both techniques moved the sample preparation step of extracting analytes to happen in situ shortly prior to ionization [11]. Control over the amount of sample surfaces wetted becomes imperative, to avoid cross-contamination and maintain lateral resolution.

In 2007, Nemes and Vertes [34] introduced laser ablation electrospray ionization (LAESI). The basic principle is the same as MALDESI, as in: ablate with laser, ionize with ESI. However, LAESI uses an IR laser and relies on water present in the sample as a makeshift matrix [35, 36] a condition most samples in live sciences fulfill. This way the deposition of an external matrix is not required, simplifying sample handling and reduces the need for manipulating the samples prior to analysis.

Very recently a combination optical microscope with laser ablation is extensively developed. It has resulted in the commercial instrument from Shimadzu, iMScope TRIO Mass microscope, combining LDI/MALDI-MS with optical microscope allowing for direct observation of the same sample using optical and mass microscopy. Further improvement in laser focusing and using liquid capture and ESI from Oak Ridge National Laboratory finally beak the 1 μm limit for MSI and structures as small as 0.5 μm could be resolved under sufficient sensitivity [37].

1.3 Array-Based Methods

Because, one key for the success of the MALDI MS-based approaches is to avoid the dilution of the analytical sample (by localizing it in a discrete micrometer-sized spot) for a more efficient laser ablation/ionization, new methods for single-cell level metabolomics analysis using MALDI-MS are based on coupling this technique with micro-dispensing devices [38], or with micro-arrays [39–43]. One example of the latter is the single-cell analytical platform known as MicroArrays for Mass Spectrometry (or MAMS). This particular type of microarrays was tailored for the study of cells grown in liquid medium using multiple-analytical read-outs, such as optical and laser desorption/ionization (LDI) or MALDI mass spectrometry. MAMS allows for the unsupervised aliquoting of cell suspensions, as demonstrated by Urban et al. in 2010 [39]. The aliquoting is achieved due to differences in wettability of the surrounding array surface and a large number of discrete micromachined recipients into this surface. These microarrays are capable of boosting the sensitivity of commercially available laser desorption/ionization (LDI) or MALDI-MS instruments by allowing the metabolic contents of the self-aliquoted cells to be localized within a discrete area (i.e., a MAMS reservoir) [39]. Furthermore, due to the transparency of the substrate used for the microarray fabrication, the whole system could be coupled to an orthogonal analytical read-out such as fluorescence [40].

MAMS has allowed the identification of co-existing phenotypes within a clonal cell culture, and observation of cell-to-cell heterogeneity in the response to stress induced by starvation [40]; the shift in the carbon source [41]; or in response to a drug [42]. One key example of the potential of microarrays when coupled to mass spectrometry was presented by Schmidt and collaborators [43]. They demonstrate the possibility of collecting single-cell level metabolic profiles using mass spectrometry and characterizing the co-existing phenotypes in an isogenic population based on the metabolic network profiles (using more than 30 metabolites detected simultaneously). The use of these networks provides systems-level information that better reflects the biological underpinnings of each phenotype, when compared to the optical detection methods.

2 Materials

The MAMS can be purchased from Sigma-Aldrich (order no. Fluka 50757), or from EMPA via a collaboration project. In the case of the latter, MAMS are fabricated as follows: transparent indium-tin oxide coated glass chips (20 mm × 20 mm × 0.7 mm) with a resistivity of 15–25 $\Omega*cm^{-1}$ were obtained commercially (Sigma-Aldrich, Buchs, Switzerland).

The slides were coated (Eposint, Pfyn, Switzerland) with a ~2–3 μm thick polysilazane coating (CAG 37, marketed by Clariant, Frankfurt am Main, Germany). This polysilazane layer was structured (EMPA, Dübendorf, Switzerland) using self-built laser ablation system equipped with a picosecond laser (SuperRapid, Lumera Laser, Kaiserslautern, Germany).

Single-cell level mass spectrometry-based metabolomics studies are challenging due to two main reasons: (a) the cells are small in size, i.e., have small sample volumes; (b) the metabolites are not equally abundant within the complex biological matrix, which pose challenges with respect to sensitivity and selectivity. Furthermore, we must consider the most suitable protocol to handle and process the cells. This is possibly the most important step for single-cell analysis because we have to process the sample in such a way that we can capture the true metabolic status of the cell [44, 45]. For example, in the case of comparative studies, it is important to efficiently and reliably remove the extracellular metabolites without compromising the integrity or modifying the metabolism of the cell [46–49]. Only if a good protocol for sample handling is employed, then we can clearly assign the source of the cell-to-cell heterogeneity without analytical artifacts even at the single-cell level [42, 43]. Therefore, we must follow all good-laboratory-practice procedures while growing and handling cell cultures. We strongly recommend that all solutions quench and process the cells, as well as those used for the MALDI matrix (if needed) to be prepared using ultrapure water and MS compatible reagents (e.g., LC–MS grade).

3 Methods

3.1 Preparing the MAMS Substrates

1. Place the MAMS substrate in a glass. Fill the container with acetone, and ultrasonicate it for 15 min at room temperature.
2. Submerge the MAMS substrate first in 50–100 mL of acetone, then in an equal volume of ethanol and finally rinse the MAMS substrate with ultrapure water.
3. Dry the MAMS substrate under a stream of nitrogen. To avoid introducing new dust particles onto the MAMS substrate use a particle filter (0.2 or 0.5 μm) in the nitrogen line.
4. Observe the MAMS substrate under a microscope to take notice of any impurity that can later be confused with the cells to be studied.

3.2 Applying the Cell Cultures onto the MAMS Substrate for Analysis

Observation: Cell handling is a very critical step in a metabolomic study. The following procedures must be carried out at low temperatures—preferably lower than −10 °C or −20 °C to keep the cells quenched—unless otherwise specified [48].

The omniphobic properties of the polymer coating used in the MAMS fabrication allow for the unsupervised aliquoting (self-aliquot) a larger liquid volume into discrete droplets (Urban et al. [39]). The MAMS microarrays are (passive) stochastic microarrays. This means that the trapping of cells onto the surface is by pulling/dragging the liquid over the MAMS surface (e.g., using a pipette tip, the MAMS Slider—Fig. 1, etc.). As a consequence, cells can be easily distributed through the MAMS substrate and a few of them

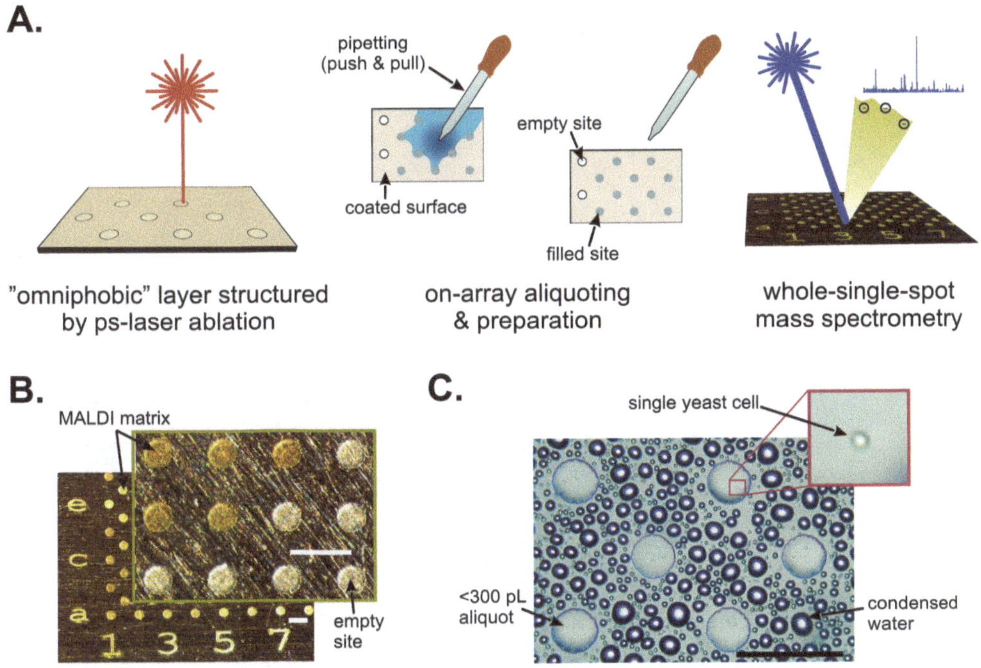

Fig. 1 Micro-arrays for mass spectrometry: (**a**) general workflow for fabrication and application of MAMS. (**b**) Homogeneous crystallization of a MALDI matrix (9-aminoacridine) aliquoted on a non-transparent MAMS with a dense arrangement of recipient sites (inset rotated by 45). The matrix solution is applied with a standard micropipette ("push and pull" method). (**c**) A transparent MAMS (based on indium tin oxide glass) useful for analysis of small cells and particles. Yeast cells (<10 mm) distributed among ~100 mm recipients; a high ambient humidity (generated with an ultrasonic humidifier). Single cells are present in some of the recipients filled with culture medium (cf. inset), while no cells are present on the adjacent coated area. Water condensation can be seen between the recipient sites. Scale bars: 300 mm (Modified from Urban et al. [15]. Copyright (2010), Royal Society of Chemistry)

will settle inside the hydrophilic wells of the MAMS substrate. It is also possible to apply the cells by using other cell handling methods such as a microspotter and microfluidic devices (Fig. 2; Amantonico et al. [38]). Since the whole cell trapping procedure is stochastic, the distribution of the number of cells within one reservoir of the MAMS substrate follows a Poisson distribution. Therefore, the density of the cell suspension applied on the MAMS substrate is directly correlated with the mean number of cells trapped in the reservoirs. It has been found experimentally that a cell density of 10^7 cells/mL provides a good number of single-cell events per MAMS substrate well.

3.3 Cell Counting

Observation: The counting of the cells can be achieved using any microscope or scanner. Once more, great care must be taken to avoid changes in the metabolite levels of the cells trap on the reservoirs (i.e., the cells must be quenched or free of any abiotic-related stress during the cell counting procedure). Once the cell

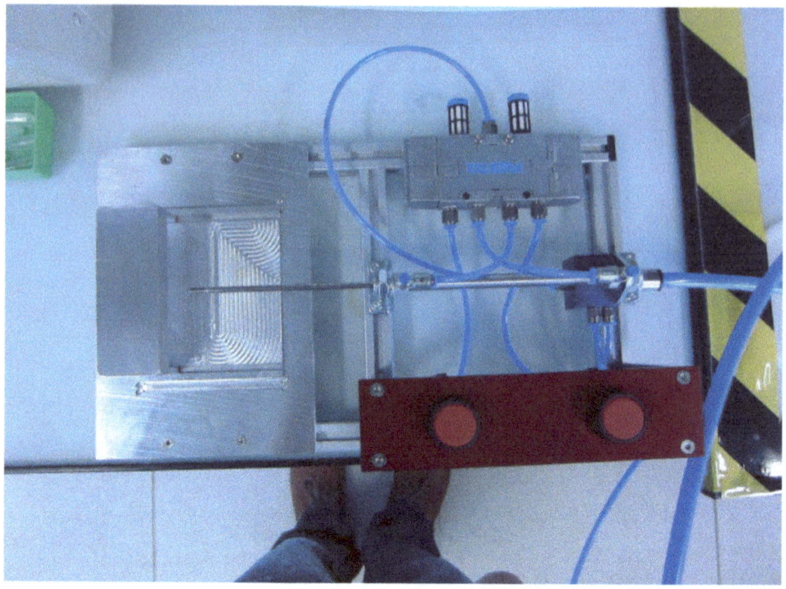

Fig. 2 MAMS slider allows for samples to be deposited by dragging one or several droplets with a metal sliding device along the MAMS lanes. By using this approach, the samples are rapidly and automatically aliquoted into the MAMS microwells due to the interplay of hydrophilic/hydrophobic interactions. With a few microliters of sample, it is possible to aliquot up to 40 replicates within seconds, since each microwell has a volume of 10 nL. More details can be found in Pabst et al. [26]

counting is completed, it is strongly suggested to freeze-dry the cells on the MAMS substrate to ensure no further enzymatic activity will occur (*See* Fig. 3 for illustration) (*See* **Note 2**).

3.4 Matrix Deposition

From an instrumental perspective, single-cell level analysis of cellular metabolites using MAMS is similar to the monitoring of metabolites by a MALDI mass spectrometry imaging (MALDI-MSI) approach (see above).

In order to use the full potential of single-cell sensitivity provided by the MAMS-based approach, an exogenous light-absorbing molecule has to be applied onto the sample (i.e., MALDI matrix). Sophisticated matrix deposition systems must be used to avoid cross-contamination between two adjacent reservoirs. Ideally, for example, it is desired that the limiting parameter for defining the distance between two adjacent hydrophilic reservoirs to be the smallest distance in which two cells can be confidently located (i.e., the laser spot size). However, several other parameters, such as (**a**) the quality of the substrate fabrication, (**b**) homogeneity of the matrix-layer, (**c**) matrix crystal size, and (**d**), most importantly, possible analyte de-localization during the matrix preparation step are having a strong influence on the

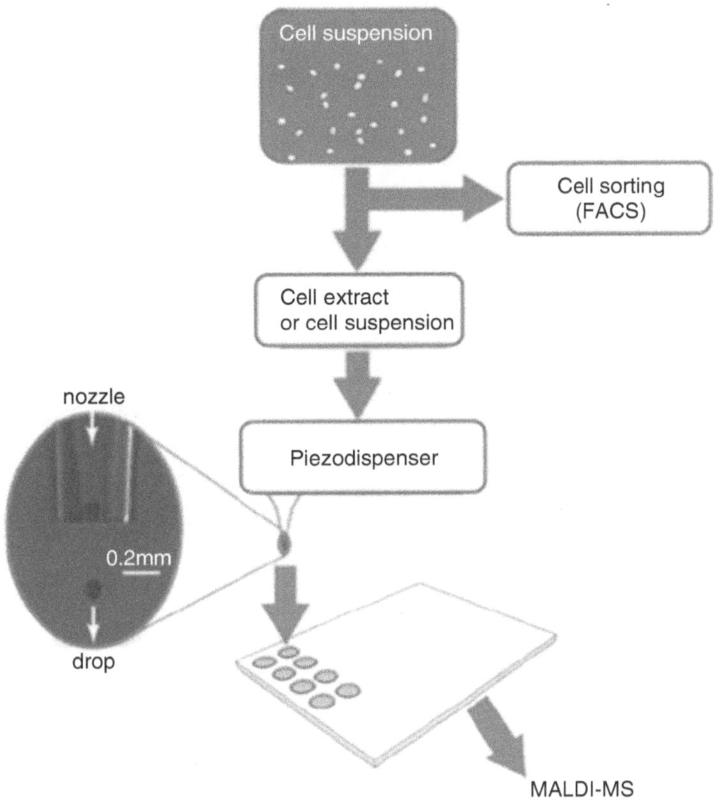

Fig. 3 Strategy to analyze the metabolites contained in a cell population. Fluorescence-activated cell sorting (FACS) was used to determine the number of cells from which the metabolites were extracted, or the number of cell phenotypes based on a reporter marker (Modified from Amantonico et al. [14]. Copyright (2008), Royal Society of Chemistry)

achievable lateral resolution, when defining the optimum distance between two adjacent reservoirs (*See* Fig. 4 for more detail) (*See* **Note 1**).

Although parameters (a, b, and c) are possible to control, the extent of analyte de-localization (d) is in many times difficult to predict. Fortunately, the advantage of the MAMS-based approach is that the analyte de-localization is minimized, since the metabolites are localized within the hydrophilic reservoirs surrounded by a hydrophobic surface.

In order to apply the MALDI matrix onto the MAMS, the MALDI matrix is typically first dissolved in a suitable solvent. Subsequently, the solvent is either spotted (e.g., using microspotters or a syringe), or sprayed (e.g., using electrosprays or pneumatic sprays, Fig. 5) during many individual drying and re-crystallization cycles onto the MAMS. An alternative technique involves the sublimation of the MALDI matrix in a vacuum chamber followed by adsorption/condensation on the sample. This method is

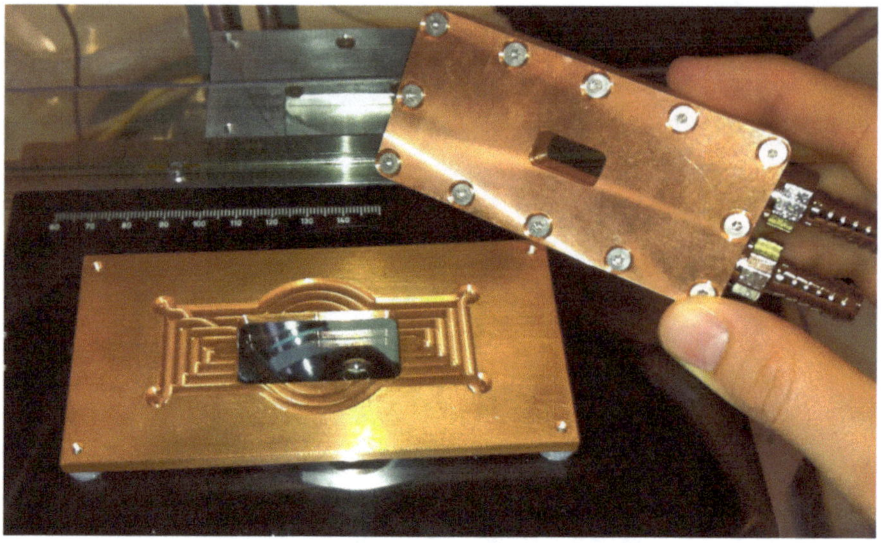

Fig. 4 In-house cryogenic stage for bright field inverse microscope. Liquid nitrogen is flushed at a constant pressure through the upper part of the stage. Metal-to-metal contact allows the temperature gradient to become homogenous. The whole stage is then placed inside a chamber that is constantly flushed with nitrogen gas. Important: A constant flush of nitrogen is applied to the glass window to avoid condensation of water

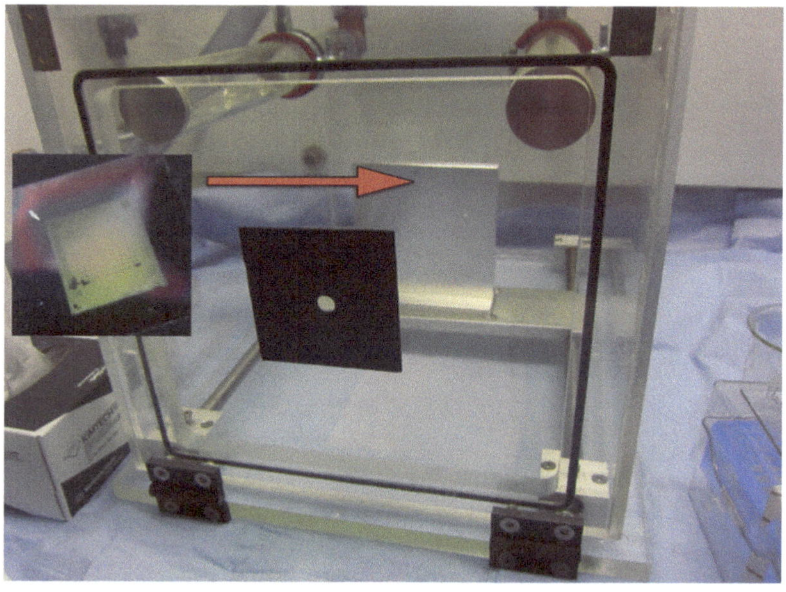

Fig. 5 Matrix deposition system developed by Dr. Ibáñez at ETH Zürich, micro-array designed and spray condition can be found in Ibanez et al. [18], or Schmidt et al. [18]

particularly attractive in MALDI-MSI since it avoids the need for solvents and was found to generate an outstanding homogeneity concerning crystal sizes and matrix coverage. However, the pure coverage of MAMS with matrix may not be sufficient for a sensitive MALDI analysis since it does not produce any matrix-analyte co-crystallization. As described above, generally an incorporation of analyte molecules into matrix crystals is necessary to obtain good MALDI-MS spectra, therefore MAMS substrates may require further treatment.

Note: Although few variants to the above-mentioned matrix application approaches have been developed to address the de-localization of metabolites on tissues as well as on the MAMS, some of which are currently marketed by vendors; there is still clearly a strong need to evaluate the efficiency of the matrix application approaches prior running a single-cell measurement to verify a good lysing of the cells and an optimum co-crystallization between analytes and MALDI matrix to ensure a sensitive analysis using MALDI instruments.

3.5 Matrix Application for Negatively Charged Molecules Using a Paint Airbrush

1. Prepare solution 9-aminoacridine (9AA) at 15 mg/mL in HPLC grade methanol or another suitable solvent that will keep the metabolism quenched.
2. Mount the MAMS substrate on a stainless steel holder plate inside a glove box for safety, *see* Fig. 5 (CAUTION, 9AA is a toxic substance, and must be handled using the appropriate gear and sprayed in well-ventilated fume hood).
3. The MAMS slides were coated using a commercial airbrush with a 0.2 mm diameter sprayer jet (airbrush tool).
4. Keep the target plate at a distance of 20 cm from the airbrush nozzle (using 100 kPa of pure nitrogen gas), and in axis with the center of the microarray. This insured that the cone of the spray reaching the target covered the entire (single-cell) microarray. Adjust the distance and the angle for different microarray sizes.
5. Spray for 20–22 s, following by 5 min drying (control the quality of the sprayed matrix visually, and adjust accordingly). The process can typically be carried out 5–10 times to provide a homogenous matrix coating. The interval between consecutive sprays insured negligible solvent build-up on the microarray, which is essential to prevent cross-contamination between the reservoirs.
6. After the last coat is applied keep the preparation under vacuum/desiccator at reduced pressure to dry it. It will shorten pumping time for the MALDI-TOF instrument and prevent MALDI source contamination.

Note: Take care that matrix coat is not pealing out after the drying. Samples with (matrix) coat damaged should not be used. Measurements should be done after reaching the appropriate vacuum without further delay to prevent extensive sublimation of 9AA into MALDI source volume and lenses.

3.6 MS Measurement of the Cells Located on the MAMS Substrate

Once the cells are on the MAMS substrate, the following steps resemble very much to MALDI MSI experiment.

1. The positions of each individual well must be introduced into the MALDI MS instrument.
2. The total material within the MAMS reservoirs must be ablated to ensure a good correlation between the MS signal and the metabolic status of the cell.

Note: There must not be a shift in signal between the first and the last reservoir to be measured due to a sublimation of matrix, water, analyte, etc. Thus control measurements on the MAMS must be performed to verify that the signal of standards aliquoted on all spots remained similar one to the other.

3.7 Data Processing and Data Analysis

To avoid erroneous molecular formula assignments, researchers are advised to follow the "Seven Golden Rules for Metabolomic Studies" that were introduced by the Fiehn Lab (Kind and Fiehn [50]). According to the "Seven Golden Rules" the identification of a metabolite must be made on the basis of accurate mass measurements, their isotope pattern, and comparing their MS/MS spectra to commercially available standards. One can find an increasing number of database entries in several databases like MassBank (http://www.massbank.jp/), NIST (http://webbook.nist.gov/chemistry/), or Metlin (https://metlin.scripps.edu/).

The spectral data are typically processed on commercial software provided with the instrument. Alternatively, *m/z* values, monoisotopic accurate mass, signal intensity, etc., can be calculated from the raw spectra using a MATLAB (MathWorks, Natick, MA, USA) peak recognition software. Normally, *m/z* values are removed with the whole spectra if the signals are not all accurately measured due to a failure in background subtraction by the software (minimum 0.001 counts in all spectra). Thus a final selection criterion to differentiate between a good from a bad spectrum is the appearance of at least 40 known metabolites. From the spectra that have been categorized as good, the raw signals are then normalized. Normalization can be done by adjusting our values with a metabolite signal that is constant in time. Another normalization procedure is the use of a linear combination of unsaturated footprint metabolite and background signals, i.e., correction factor = α α, β,. . ., ω × [Signal A] +, β × [Signal B] + . . . ω ω × [Signal X]. The unsaturated signals were chosen because they did not correlate with the signals of biological origin; while the weight values (α, β,. . ., ω) were empirically calculated (*See* **Note 4**).

3.8 Control Measurements for Analytical Validation

To demonstrate that our MAMS-based measurements are actually monitoring biological information and t are not influenced by instrumental and sample handling variation, it is necessary to compare the correlation between two, three, or more metabolites.

4 Notes

1. Independently of the cell type used in our study, sample preparation (i.e., cell handling and processing) is possibly the most problematic step for single-cell analysis, in particular using the MAMS substrate. For example, due to the low temperature requirements to quench cell metabolism during cell handling, cells will be more commonly suspended in organic solvents. Unfortunately, the use of organic solvents enables metabolites to leak from cells, which translate in the loss of sensitivity for some metabolites. Thus, it is highly recommended that scientists use quenching protocols that they have previously used in population-level studies and showed no analytical artifacts. In other words, the researchers are sure that the metabolites of interest are only intracellular metabolites (and not extracellular metabolites found in the cell suspension), and that the metabolite extraction will mainly happen during the co-crystallization of the cells with the MALDI matrix.
2. Freeze-drying is a dehydration process. It works by decreasing the temperature of the material for example by flowing liquid nitrogen around the chamber that contains the sample. Then, the surrounding air is removed using vacuum pumps. At reaching proper vacuum, the frozen water in the material starts sublimating thus drying the material in the process.
3. The key point for a good MALDI matrix application is the optimum co-crystallization between the endogenous cell metabolites and the MALDI matrix, while at the same time, trying to avoid any cross-contamination, or de-localization of the same metabolites. (In common words, the researcher must be able to clearly identify the origin of the mass spectrometry signal, and associate this signal to the cell (or cells) located within a single MAMS reservoir.)
4. The key for data processing is to differentiate between footprint and fingerprint metabolites (i.e., metabolites that are exogenous or endogenous to the cells respectively). Dependent on the cell concentration employed, the number of cells on the substrate will follow a Poisson distribution of random events. Thus, we can differentiate between the metabolites in the culture and washing media (footprinting) and the metabolites

within the cell (fingerprinting) by observing a correlation between the number of cells and the signal intensity of the metabolites.

References

1. Ralser M, Wamelink MM, Kowald A, Gerisch B, Heeren G, Struys EA, Klipp E, Jakobs C, Breitenbach M, Lehrach H, Krobitsch S (2007) Dynamic rerouting of the carbohydrate flux is key to counteracting oxidative stress. J Biol 6:10
2. Sauer U, Zamboni N (2008) From biomarkers to integrated network responses. Nat Biotechnol 26:1090–1092
3. Novick A, Weiner M (1957) Enzyme induction as an all-or-none phenomenon. Proc Natl Acad Sci U S A 43:553–566
4. Rosenbluth MJ, Lam WA, Fletcher DA (2008) Analyzing cell mechanics in hematologic diseases with microfluidic biophysical flow cytometry. Lab Chip 8:1062–1070
5. Balaban NQ, Merrin J, Chait R, Kowalik L, Leibler S (2004) Bacterial persistence as a phenotypic switch. Science 305:1622–1625
6. Altschuler SJ, Wu LF (2010) Cellular heterogeneity: do differences make a difference? Cell 141:559–563
7. Raj A, van Oudenaarden A (2008) Nature, nurture, or chance: stochastic gene expression and its consequences. Cell 135:216–226
8. Avery SV (2006) Microbial cell individuality and the underlying sources of heterogeneity. Nat Rev Microbiol 4(8):577–587
9. Bhardwaj C, Hanley L (2014) Ion sources for mass spectrometric identification and imaging of molecular species. Nat Prod Rep 31:756–767
10. El-Baba TJ, Lutomski CA, Wang BX, Inutan ED, Trimpin S (2014) Toward high spatial resolution sampling and characterization of biological tissue surfaces using mass spectrometry. Anal Bioanal Chem 406:4053–4061
11. Venter AR, Douglass KA, Shelley JT, Hasman G, Honarvar E (2014) Mechanisms of real-time, proximal sample processing during ambient ionization mass spectrometry. Anal Chem 86:233–249
12. Bartels B, Svatoš A (2015) Spatially resolved in vivo plant metabolomics by laser ablation-based mass spectrometry imaging (MSI) techniques: LDI-MSI and LAESI. Front Plant Sci 6:471
13. Shrivas K, Setou M (2012) Imaging mass spectrometry: sample preparation, instrumentation, and applications. In: Hawkes PW (ed) Advances in imaging and electron physics, vol 171. Elsevier, Oxford, pp 145–193
14. McMahon JM, Dookeran NN, Todd PJ (1995) Organic ion imaging beyond the limit of static secondary-ion mass-spectrometry. J Am Soc Mass Spectrom 6:1047–1058
15. Caprioli RM, Farmer TB, Gile J (1997) Molecular imaging of biological samples: localization of peptides and proteins using MALDI-TOF MS. Anal Chem 69:4751–4760
16. Svatos A (2010) Mass spectrometric imaging of small molecules. Trends Biotechnol 28:425–434
17. Svatoš A (2011) Single-cell metabolomics comes of age: new developments in mass spectrometry profiling and imaging. Anal Chem 83:5037–5044
18. Li H, Smith BK, Shrestha B, Mark L, Vertes A (2015) Automated cell-by-cell tissue imaging and single-cell analysis for targeted morphologies by laser ablation electrospray ionization mass spectrometry. In: He L (ed) Mass spectrometry imaging of small molecules. Humana Press, New York, pp 117–127
19. Feigl P, Schueler B, Hillenkamp F (1983) LAMMA-1000, a new instrument for bulk microprobe mass analysis by pulsed laser irradiation. Int J Mass Spectrom Ion Process 47:15–18
20. Karas M, Bachmann D, Hillenkamp F (1985) Influence of the wavelength in high-irradiance ultraviolet-laser desorption mass-spectrometry of organic-molecules. Anal Chem 57:2935–2939
21. Karas M, Bachmann D, Bahr U, Hillenkamp F (1987) Matrix-assisted ultraviolet-laser desorption of nonvolatile compounds. Int J Mass Spectrom Ion Process 78:53–68
22. Karas M, Kruger R (2003) Ion formation in MALDI: the cluster ionization mechanism. Chem Rev 103:427–439
23. Laiko VV, Baldwin MA, Burlingame AL (2000) Atmospheric pressure matrix assisted laser desorption/ionization mass spectrometry. Anal Chem 72:652–657
24. Li Y, Shrestha B, Vertes A (2007) Atmospheric pressure molecular imaging by infrared MALDI mass spectrometry. Anal Chem 79:523–532

25. Hoelscher D, Shroff R, Knop K, Gottschaldt M, Crecelius A, Schneider B, Heckel DG, Schubert US, Svatos A (2009) Matrix-free UV-laser desorption/ionization (LDI) mass spectrometric imaging at the single-cell level: distribution of secondary metabolites of *Arabidopsis thaliana* and Hypericum species. Plant J 60:907–918
26. Schober Y, Guenther S, Spengler B, Rompp A (2012) Single cell matrix-assisted laser desorption/ionization mass spectrometry imaging. Anal Chem 84:6293–6297
27. Becker JS, Zoriy M, Matusch A, Wu B, Salber D, Palm C (2010) Bioimaging of metals by laser ablation inductively coupled plasma mass spectrometry (LA-ICP-MS). Mass Spectrom Rev 29:156–175
28. Bendall SC, Simonds EF, Qiu P, Amir EaD, Krutzik PO, Finck R, Bruggner RV, Melamed R, Trejo A, Ornatsky OI, Balderas RS, Plevritis SK, Sachs K, Pe'er D, Tanner SD, Nolan GP (2011) Single-cell mass cytometry of differential immune and drug responses across a human hematopoietic continuum. Science 332:687–696
29. Sampson JS, Hawkridge AM, Muddiman DC (2006) Generation and detection of multiply-charged peptides and proteins by matrix-assisted laser desorption electrospray ionization (MALDESI) Fourier transform ion cyclotron resonance mass spectrometry. J Am Soc Mass Spectrom 17:1712–1716
30. Robichaud G, Barry JA, Muddiman DC (2014) IR-MALDESI mass spectrometry imaging of biological tissue sections using ice as a matrix. J Am Soc Mass Spectrom 25:319–328
31. Bjarnholt N, Li B, D'alvise J, Janfelt C (2014) Mass spectrometry imaging of plant metabolites - principles and possibilities. Nat Prod Rep 31:818–837
32. Lanekoff I, Heath BS, Liyu A, Thomas M, Carson JP, Laskin J (2012) Automated platform for high-resolution tissue imaging using Nanospray desorption electrospray ionization mass spectrometry. Anal Chem 84:8351–8356
33. Campbell DI, Ferreira CR, Eberlin LS, Cooks RG (2012) Improved spatial resolution in the imaging of biological tissue using desorption electrospray ionization. Anal Bioanal Chem 404:389–398
34. Nemes P, Vertes A (2007) Laser ablation electrospray ionization for atmospheric pressure, in vivo, and imaging mass spectrometry. Anal Chem 79:8098–8106
35. Apitz I, Vogel A (2005) Material ejection in nanosecond Er: YAG laser ablation of water, liver, and skin. Appl Phys A 81:329–338
36. Nemes P, Huang HH, Vertes A (2012) Internal energy deposition and ion fragmentation in atmospheric-pressure mid-infrared laser ablation electrospray ionization. Phys Chem Chem Phys 14:2501–2507
37. Cahill JF, Kertesz V, Van Berkel GJ (2015) Characterization and application of a hybrid optical microscopy/laser ablation liquid vortex capture/electrospray ionization system for mass spectrometry imaging with sub-micrometer spatial resolution. Anal Chem 87:11113–11121
38. Amantonico A, Oh JY, Sobek J, Heinemann M, Zenobi R (2008) Mass spectrometric method for analyzing metabolites in yeast with single cell sensitivity. Angew Chem Int Ed Engl 47(29):5382–5385
39. Urban PL, Jefimovs K, Amantonico A, Fagerer SR, Schmid T, Mädler S, Puigmarti-Luis J, Goedecke N, Zenobi R (2010) High-density micro-arrays for mass spectrometry. Lab Chip 10(23):3206–3209
40. Fagerer SR, Schmid T, Ibáñez AJ, Pabst M, Steinhoff R, Jefimovs K, Urban PL, Zenobi R (2013) Analysis of single algal cells by combining mass spectrometry with Raman and fluorescence mapping. Analyst 138:6732–6736
41. Urban PL, Schmidt AM, Fagerer SR, Amantonico A, Ibañez AJ, Jefimovs K, Heinemann M, Zenobi R (2011) Carbon-13 labelling strategy for studying the ATP metabolism in individual yeast cells by micro-arrays for mass spectrometry. Mol BioSyst 7 (10):2837–2840
42. Ibáñez AJ, Fagerer SR, Schmidt AM, Urban PL, Jefimovs K, Geiger P, Dechant R, Heinemann M, Zenobi R (2013) Mass spectrometry-based metabolomics of single yeast cells. Proc Natl Acad Sci U S A 110 (22):8790–8794
43. Schmidt AM, Fagerer SR, Jefimovs K, Buettner F, Marro C, Siringil EC, Boehlen KL, Pabst M, Ibáñez AJ (2014) Molecular phenotypic profiling of a Saccharomyces cerevisiae strain at the single-cell level. Analyst 139 (22):5709–5717
44. Kresnowati MT, van Winden WA, Almering MJ, ten Pierick A, Ras C, Knijnenburg TA, Daran-Lapujade P, Pronk JT, Heijnen JJ, Daran JM (2006) When transcriptome meets metabolome: fast cellular responses of yeast to

sudden relief of glucose limitation. Mol Syst Biol 2:49

45. Brauer MJ, Yuan J, Bennett BD, Lu W, Kimball E, Botstein D, Rabinowitz JD (2006) Conservation of the metabolomic response to starvation across two divergent microbes. Proc Natl Acad Sci U S A 103:19302–19307
46. Winder CL, Dunn WB, Schuler S, Broadhurst D, Jarvis R, Stephens GM, Goodacre R (2008) Global metabolic profiling of Escherichia coli cultures: an evaluation of methods for quenching and extraction of intracellular metabolites. Anal Chem 80:2939–2948
47. Faijes M, Mars AE, Smid EJ (2007) Comparison of quenching and extraction methodologies for metabolome analysis of Lactobacillus plantarum. Microb Cell Fact 6:27
48. Sellick CA, Hansen R, Stephens GM, Goodacre R, Dickson AJ (2011) Metabolite extraction from suspension-cultured mammalian cells for global metabolite profiling. Nat Protoc 6:1241–1249
49. Canelas AB, Ras C, ten Pierick A, van Dam JC, Hiejnen JJ, van Gulik WM (2008) Leakage-free rapid quenching technique for yeast metabolomics. Metabolomics 4:226–239
50. Kind T, Fiehn O (2007) Seven Golden Rules for heuristic filtering of molecular formulas obtained by accurate mass spectrometry. BMC Bioinformatics 8:105

Chapter 7

Laser Capture Microdissection–Liquid Vortex Capture Mass Spectrometry Metabolic Profiling of Single Onion Epidermis and Microalgae Cells

John F. Cahill and Vilmos Kertesz

Abstract

Laser capture microdissection is a valuable technique in individually isolating single cells whether in tissue networks or deposited from a cell suspension. New developments have enabled coupling of laser capture microdissection with mass spectrometry via liquid vortex capture sampling probe. This enables online metabolic profiling of sectioned cells. Here, we describe the protocol used to deposit, isolate, and individually chemically characterize single *Allium cepa* and *Chlamydomonas reinhardtii* cells by laser capture microdissection–liquid vortex capture mass spectrometry.

Key words Laser capture microdissection, Mass spectrometry, Single cell, Onion epidermis cell, Microalgae, Liquid-vortex capture, Metabolites

1 Introduction

Cells are naturally heterogeneous and have differences in chemistry due to their history, age, and microenvironment, among other factors [1, 2]. Knowledge of the metabolomic and/or lipidomic state of each individual cell promises to yield unprecedented insight into cellular mechanisms. However, chemical analysis of single cells is challenging as individual cells contain very little material but are also chemically complex, containing millions of discrete molecular species. Developments in mass spectrometry, in particular, atmospheric pressure sampling methodologies, have begun to enable chemical measurement of single cells in an untargeted manner [3–7].

One such technique is laser capture microdissection–liquid vortex capture mass spectrometry (LMD-LVC-MS) [5, 6, 8]. In this technique, an LMD device is used to laser ablate or to carefully section sample material which is subsequently captured by an LVC-MS sampling probe. Once the material is captured into the

Bindesh Shrestha (ed.), *Single Cell Metabolism: Methods and Protocols*, Methods in Molecular Biology, vol. 2064,
https://doi.org/10.1007/978-1-4939-9831-9_7,

solvent of the probe, soluble analytes are extracted and transported to the ionization source of the mass spectrometer. From there, the analyte molecules are ionized and measured by mass spectrometry. One of the unique capabilities of the LMD-LVC-MS technique is the ability to isolate and measure the chemistry of single cells. In this chapter, we describe a method to acquire single cell mass spectra from plant thin tissue and microalgae cell in suspension using LMD-LVC-MS. In the present case, we use *Allium cepa* and *Chlamydomonas reinhardtii* cells, but the protocol is applicable to many plants (single-cell-thick tissue) and microalgae.

2 Materials

2.1 Chemicals

1. HPLC-grade water.
2. HPLC-grade methanol.
3. Beijerinck's solution (Chlamydomonas Resource Center, St. Paul, MN, USA).
4. Phosphate solution (Chlamydomonas Resource Center, St. Paul, MN, USA).
5. Hutner's trace elements (Chlamydomonas Resource Center, St. Paul, MN, USA).

2.2 Plant

1. *Allium cepa* (purchased locally).

2.3 Microalgae

1. CC-125 wild type mt+ [137c] (Chlamydomonas Resource Center, St. Paul, MN, USA).

2.4 Equipment

1. Syringe pump (Harvard apparatus, Holliston, MA, USA).
2. Syringes.
3. Pipettes.
4. Tweezers.
5. Razor blade or sharp knife.
6. Centrifuge (Eppendorf 5430 centrifuge, Hamburg, Germany).
7. Heat shrink tubing.
8. Large plastic sheet.
9. Blue sharpie.
10. Teflon tubing (IDEX Health & Science LLC, Oak Harbor, WA, USA).
11. FEP sleave (0.794 mm-i.d. × 1.59 mm-o.d., IDEX Health & Science LLC, Oak Harbor, WA, USA).
12. Stainless steel tubing 1.28 mm-i.d. × ~1.61 mm-o.d. (McMaster-Carr, Douglasville, GA, USA).

13. 30-gauge miniature stainless steel tubing (McMaster-Carr, Douglasville, GA, USA).
14. Standard 1″ × 3″ microscope glass slides (we used Fisherbrand™ Superfrost™ precleaned slides, Fisher Scientific, Waltham, MA, USA).
15. 2.0 μm PEN membrane slides (P/N: 11505158, Leica Microsystems, Wetzlar, Germany).
16. 5-Port PEEK manifold assembly (IDEX Health & Science LLC, Oak Harbor, WA, USA).
17. Flangeless fittings (IDEX Health & Science LLC, Oak Harbor, WA, USA).
18. 1/8″ and 1/16″ ferrules (IDEX Health & Science LLC, Oak Harbor, WA, USA).
19. 178 μm-i.d. × 794 μm-o.d. × ~24-cm-long PEEK capillary (IDEX Health & Science LLC, Oak Harbor, WA, USA).
20. Mass spectrometer (TripleTOF® 5600+ from SCIEX, Concord, Ontario, Canada)
21. Laser capture microdissection system (LMD7000 from Leica Microsystems, Wetzlar, Germany).
22. CCD camera (Costar SI-C600N, Coppell, TX, USA).
23. Lift table (GS BS75, Global Industrial, Port Washington, NY, USA).
24. 50 mL plastic centrifuge tubes.
25. White fluorescent light.
26. Orbital shaker (Helix 150BLR orbital shaker, LabStrong, Dubuque, IA, USA).

3 Methods

3.1 Onion Skin Single Cell Epidermal Slide Preparation

1. Peel off the outmost layer of onion and remove one inner scale.
2. Using a razor blade (or a sharp knife) cut out a ~0.5″ × 1″ section.
3. Put your thumbs on the outside of the section and carefully snap that in half such that the epidermis layer on the smooth/shiny side stays intact.
4. Very gently pull the two pieces apart. This separates the epidermis layer (thin, transparent layer of cells) from the underlying layer.
5. When the epidermis layer is separated from one half of the section, use tweezers to grab the epidermis layer and gently separate it from the other half.

6. Place the epidermis layer on the glass microscope slide and flatten it.
7. Let the epidermis layer air dry for about 30 min (*see* **Note 1**).

3.2 Microalgae Slide Preparation

1. Prepare High Salt Medium (HSM) by mixing 5 mL Beijerinck's solution, 5 mL phosphate solution, and 1 mL of Hutner's trace elements with 1 L of water.
2. Culture CC125 in ~15 mL of HSM media in a 50 mL centrifuge tube under constant shaking, light, and temperature (25 °C). The cell culture will take several days to generate a significant number of cells. The culture will have a deep green color. A hemacytometer can be used to determine cell number concentration.
3. Prepare microalgae samples on microscope slides by pipetting 10 μL of cell solution directly onto PEN membrane slide and let air dry. Repeat as necessary (*see* **Note 2**).

3.3 LMD-LVC-MS Setup

1. The LVC-MS probe used for these experiments was identical to the low-profile LVC probe that was used in Cahill et al. (Fig. 1) [8]. The LVC probe consists of a coaxial tube design. The outer tube was 1.28 mm-i.d. × ~1.61 mm-o.d. × 1.1 cm long stainless steel, and the inner was a 178 μm-i.d. × 794 μm-o.d. × ~24-cm-long PEEK capillary (*see* **Note 3**). To keep the inner PEEK capillary centered in the outer tube

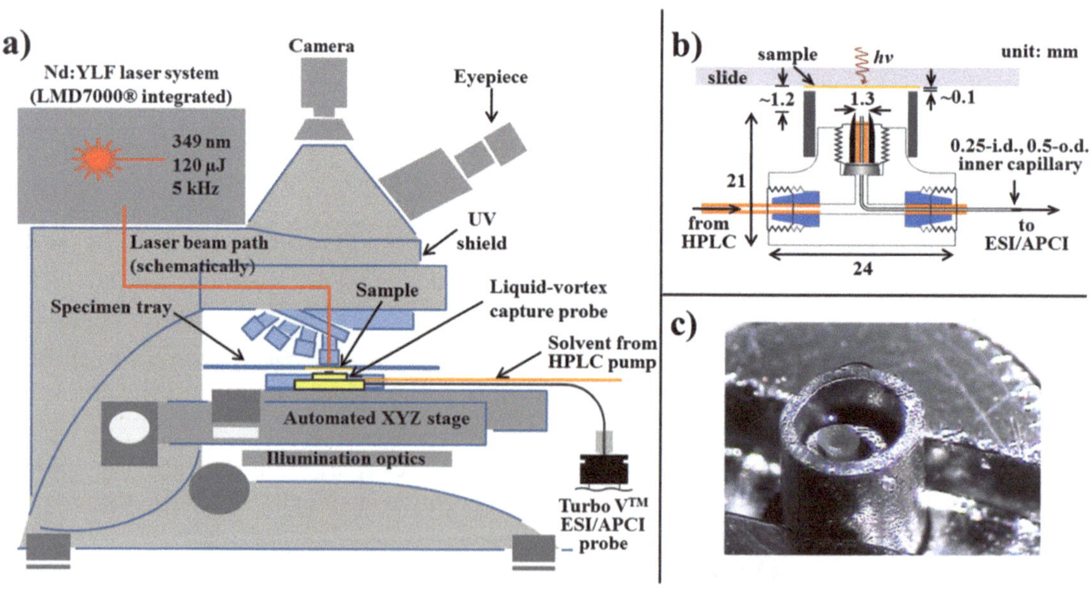

Fig. 1 (**a**) Experimental design of the coupled LMD7000-LVC mass spectrometry system. (**b**) Schematic of the LVC probe design and positioning relative to the sample. (**c**) Photograph of the top of the LVC probe (Reproduced from [8] with permission from the American Chemical Society)

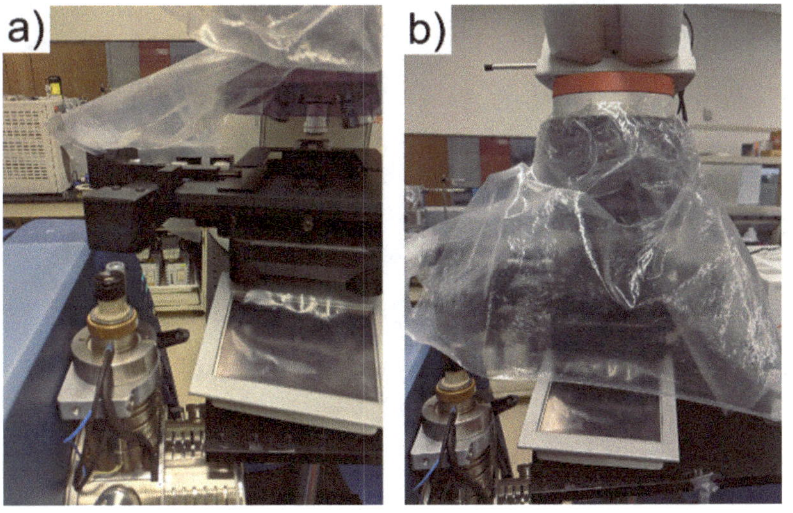

Fig. 2 (**a**) Positioning of the LMD system to the inlet of the ionization source of the mass spectrometer. (**b**) Plastic covering used to block negative airflows around the LMD system

we used a short FEP sleave (0.794 mm-i.d. × 1.59 mm-o.d.). The outer tube was connected to the mass spectrometer ion source electrical ground via a metal wire. Each tube was secured to a 5-port PEEK manifold that was machined down from its original 31 mm circular base diameter to a 24 mm square shape. Two of the four horizontal ports of the manifold were plugged. Solvent was delivered through one of the remaining horizontal ports by an HPLC pump into the annulus region of the two coaxial tubes mounted in the vertical port of the manifold. The PEEK capillary went through the last remaining port of the manifold and eventually connected to the ion source of the mass spectrometer (*see* **Note 4**).

2. Replace the capillary of the mass spectrometer ion source with a 30-gauge stainless steel capillary of the same length. This will allow greater solvent flow rates for the LVC probe.
3. Position and arrange the LMD system on a lift table to get as close as possible to the inlet of the ion source of the mass spectrometer (*see* Fig. 2a). *Caution*: Poor positioning can cause the LMD system to collide with the ion source of the mass spectrometer! Be careful to provide enough room so that the initialization procedure and movement of the LMD system while focusing will not cause the system to collide with the mass spectrometer.
4. Remove the sample shield (metal plate directly under the sample slide holder) from the LMD system by removing two screws on the outside of the system.

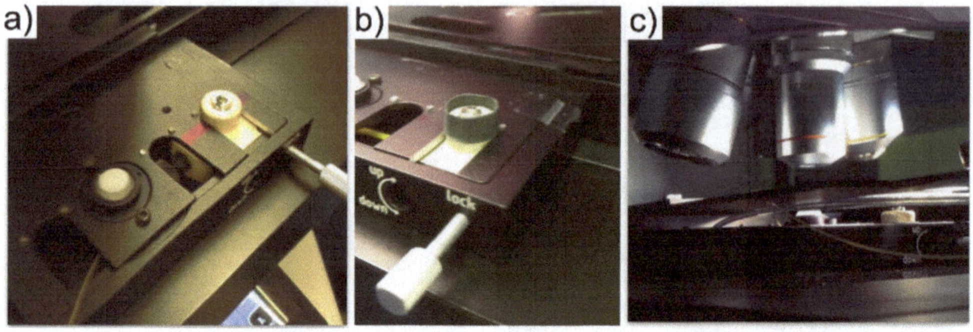

Fig. 3 (**a**) LVC probe in the 'D' slot of the disassembled collector. (**b**) LVC probe with the heat shrink sheath. (**c**) LVC probe positioned under the sample holder. Note the heat shrink sheath was removed in (**c**) for visual clarity

5. With the LMD on and initialized, press the 'Unload Collector' button to allow access to the sample collector.
6. Disassemble the LMD collector by removing the two screws on top. Mount the LVC probe into the 'D' slot of the bottom piece of the disassembled collector (*see* Fig. 3a).
7. Put the collector with the LVC probe back into the LMD device (*see* **Note 5**). Remove the glass slide holder. Verify that the top of the LVC probe is below the sample slide holder cap before re-insertion of the LMD capture cap device. *Caution*: If LVC probe is too tall it can collide with the sample slide holder!
8. Align the center of the LVC probe under the laser focus of the microscope by adjusting the reference point position of the collector in the LMD system. To adjust press the 'Unload Collector' button to access the reference point position controls. Adjust x/y positions and load the collector back into position. Move the collector to the 'D' position and check the alignment of the microscope objective with the center of the LVC probe. Repeat as necessary to fine tune the reference point position. In this case a setting of $x = 36.3$ mm and $y = 5.3$ mm was used (*see* **Note 6**).
9. Block negative airflows by covering the laser ablation area in a plastic sheet (Fig. 2b). We used a large plastic bag taped to the UV shield of the LMD system. Use a 12-mm-long FEP sleave (heat shrink tubing) contoured to the top of the 5-port PEEK manifold to also reduce the influence of negative airflows (Fig. 3b). The top of the heat shrink tubing should extend ~0.5 mm above the top of the LVC probe (*see* **Note 7**).
10. Connect the end of LVC inner capillary to the ion source of the mass spectrometer. Turn mass spectrometer gas flow settings to maximum. In our setup a GS1/GS2 flows were 90/60.

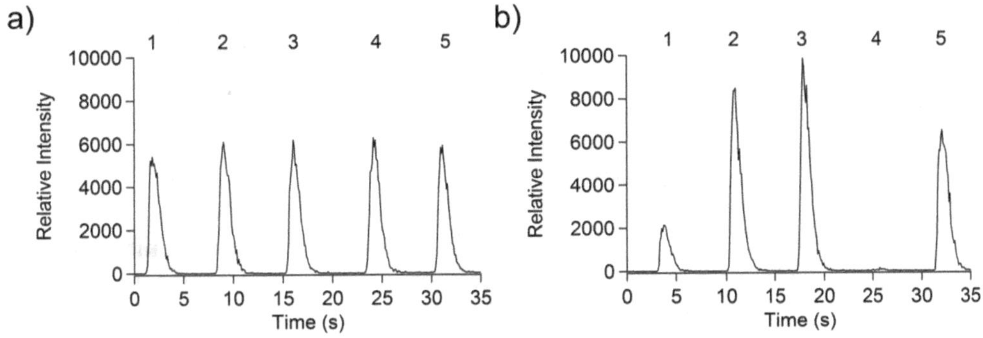

Fig. 4 Extracted ion chromatograms from 5, ~1.9-μm-diameter laser ablation spots of Basic Blue 7 containing inked films. Instrument parameters: *m/z* 478 → 434, IS = 5.5 kV, DP = 35 V, CE = 50 eV, 0.150 mL/min 80/20 methanol/water +0.1% formic acid (Reproduced from [8] with permission from the American Chemical Society)

11. Pump in solvent from the HPLC or syringe pump. Tune the solvent flow rate to make a stable vortex in the LVC probe (*see* Fig. 1c and **Note 8**). In our setup a solvent flow of 150 μL/min 100/0.1 (v/v) methanol/formic acid created a stable vortex (*see* **Note 9**).
12. Optimize the LMD-LVC-MS setup: Mark a glass slide with blue Sharpie marker and set mass spectrometer for selected reaction monitoring (SRM) of the Basic Blue 7 dye transition (*m/z* 478 → 434, ion spray (IS) voltage = 5.5 kV) (*see* **Note 10**), heated nebulizer temperature (TEM) = 450 °C, declustering potential (DP) = 35 V, collision energy (CE) = 50 eV, 150 μL/min 80/20/0.1 (v/v/v) methanol/water/formic acid. Move collection cap to 'D' position and ablate the blue Sharpie using the 'Move and Cut' option for laser ablation. Optimize solvent flow rate and LVC probe position to obtain the best signal profile (*see* Fig. 4 and **Note 11**).

3.4 LMD-LVC-MS Sampling

1. For onion tissue place the glass slide with dried onion epidermis layer facing down on the sample holder (i.e., facing the LVC probe). Find onion epidermis cell with the microscope. Select one cell to sample and note its position.
2. For algae cells deposited from suspension place the PEN membrane slide with dried microalgae facing down on the sample holder (i.e., facing the LVC probe). Find algae with the microscope. Select one cell to sample and note its position (*see* **Note 12**).
3. Optimize mass spectrometer settings and SRM transitions for the compounds you wish to monitor and the LVC solvent composition and flowrate for vortexing. In this setup, a solvent flow of 0.150 mL/min 100/0.1 (v/v) methanol/formic

acid was used. For algae, SRM scans were employed to monitor for monogalactosyldiacylglycerol (MGDG) and diacylglyceryltrimethylhomo-Ser (DGTS) lipids. MS parameters optimized for each lipid were: MGDG: *m/z* 762.7 → 335.3, IS = 5.5 kV, TEM = 450 °C, DP = 100 V, CE =30 V and DGTS: *m/z* 732.6 → 236.2, IS = 5.5 kV, TEM = 450 °C, DP = 100 V, CE =65 V. For onion epidermis cells, mono-, di-, and trisaccharides were monitored using full scan time-of-flight mass spectra in positive ion mode with optimized instrument parameters (IS = 5.5 kV, TEM = 450 °C, DP = 100 V).

4. Choose which laser sampling method to use: laser ablation spot sampling (algae), laser ablation raster sampling (algae or onion epidermis), or 'Cut and Drop' sampling (algae).

5. Laser ablation spot sampling.

   (a) Select 'Move and Cut' option for laser ablation in the laser settings menu. Adjust laser power and focus to make laser spot size greater than the cell (~10 μm for *Chlamydomonas reinhardtii* microalgae). Move collection cap to 'D' position. This moves the LVC probe directly under the cell. Point mouse cursor over the cell and click to fire the laser and ablate the cell. Single cell sampling of *Chlamydomonas reinhardtii* is shown in Fig. 5.

6. Laser raster sampling.

   (a) Draw a custom shape outlining the cell. For *Chlamydomonas reinhardtii* a circle works well. For *Allium cepa* a custom elliptical shape (~150 μm in length) is generally needed. After shape is created select capture cap 'D' in the 'Collecting Tube Cap' prompt. Select the 'Draw and Scan' option for laser ablation in the laser settings menu and adjust laser power and scan rate options as desired (*see* **Note 13**). Click 'Scan' to laser raster the cell. Figure 6 shows exemplary data acquired from *Allium cepa* cells.

7. 'Cut and Drop' sampling.

   (a) Draw a custom shape encompassing the cell. Leave a small gap around the outside of the cell (>spot size of laser). After shape is created select capture cap 'D' in the prompt. Click 'Draw and Cut' to laser ablate around the cell (*see* **Note 14**). If the section is still suspended select 'Point and cut' option for laser cutting and point and click inside the shape to eject the cell microdissection (*see* **Note 15**). Examples of 'Cut and Drop' sampling of cells and tissue can be found in references [5, 6, 9, 10].

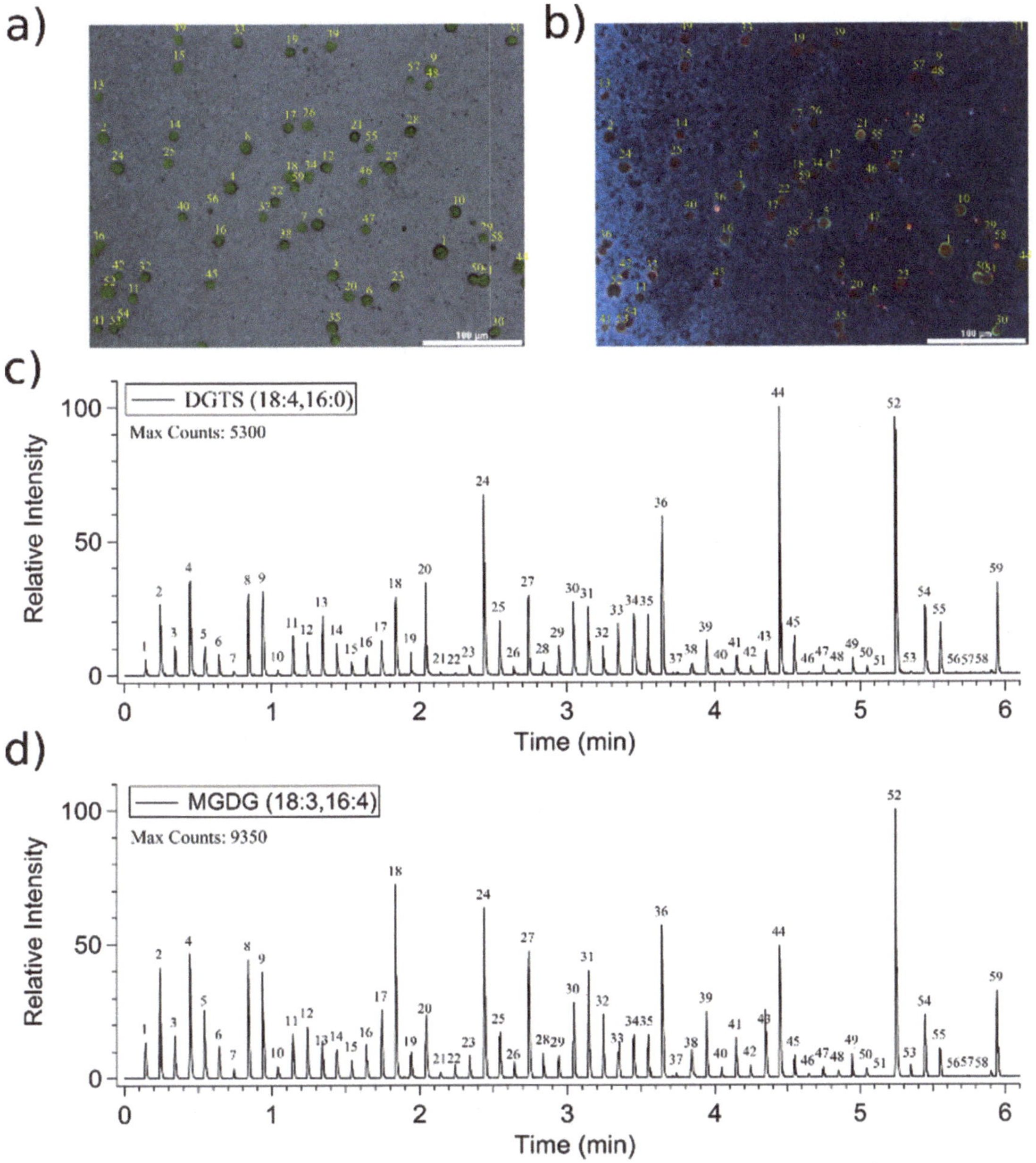

Fig. 5 (**a**) Bright-field and (**b**) fluorescence images of *Chlamydomonas reinhardtii* cells deposited onto a glass slide. Extracted ion chromatograms of (**c**) DGTS (18:4, 16:0) and (**d**) MGDG (18:3, 16:4) lipids. Numbers above peaks in (**c**) and (**d**) reflect the corresponding cells in (**a**) and (**b**). Image scale bars are 100 μm. (Reproduced from [6] with permission from John Wiley & Sons Ltd.)

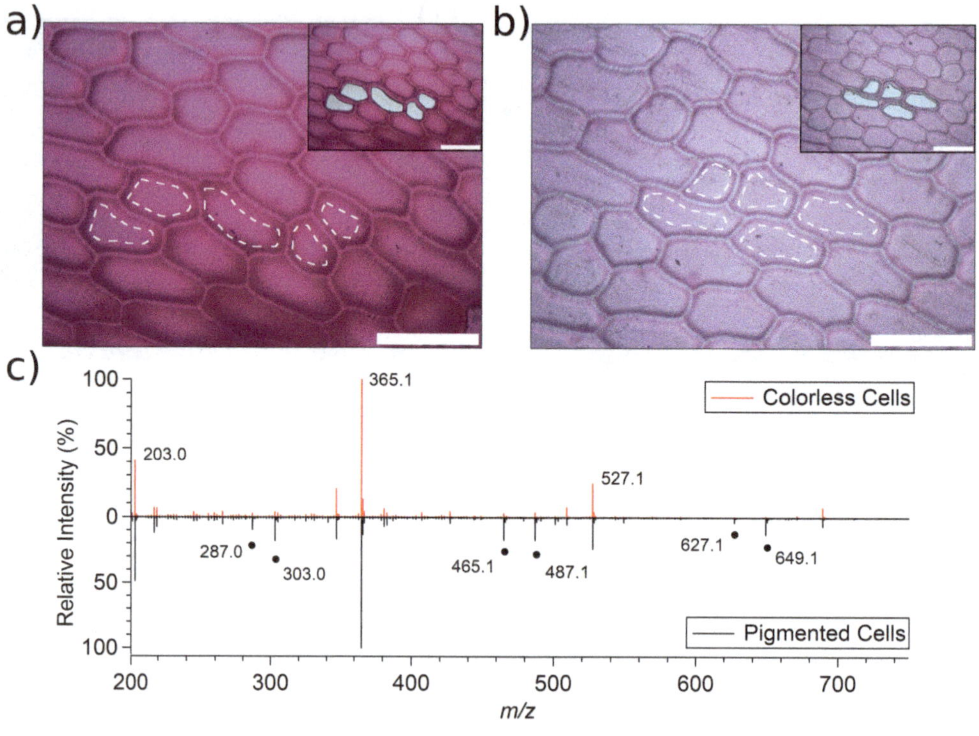

Fig. 6 Bright-field images of (**a**) pigmented and (**b**) colorless *Allium cepa* epidermis cells. For each cell the exact shape was drawn in LMD software and then sampled by quickly rastering the laser within the shape (white dotted lines). Insets show cells after laser ablation. (**c**) Average mass spectra of all pigmented (black line) and colorless (red line) cells. Black dots indicate a number of ions unique to pigmented cells. Scale bars are 200 μm (Reproduced from [6] with permission from John Wiley & Sons Ltd.)

4 Notes

1. This video, https://www.youtube.com/watch?v=PrX3h-AflZI, also shows how to get a single cell layer from the onion epidermis.
2. Algae can be deposited directly from HSM solution but if too much salt is present either dilute in water or centrifuge the cells for 5 min at 600 × *g*, remove supernatant, and resuspend in a 50/50 (v/v) HSM/water mixture.
3. The exact length of the PEEK capillary will depend on final positioning of LMD system in relation to the mass spectrometer. Thus, it is practical to construct the LVC probe with excess PEEK capillary length (~30 cm) and then trim to a shorter length once the LMD system is in position.
4. The PEEK capillary will have to make a near 90° turn. To achieve this a small tube inserted in the opposite horizontal port can be used to push the PEEK capillary through. Damaged areas of the capillary should be trimmed off.

5. The solvent delivery and inner capillaries tend to get caught on the stage during movement. To avoid this simply tape the capillary flat against the collection cap holder side.
6. Fluorescence imaging can be used to view the exact position of the LVC probe under the microscope. The PEEK capillary will fluoresce brightly.
7. Our setup was near to the air conditioning vent in the laboratory which created some significant airflow around the LMD system. Positioning of the system (and mass spectrometer) away from possibly detrimental airflows may negate the need for the plastic covering and heat shrink sheath. The heat shrink sheath should extend to just barely touching the sample holder. Positioning the heat shrink sheath to be slightly higher than the sample holder height and then moving the collector into position under the sample holder would push the heat shrink sheath into an optimal position.
8. If LVC vortex is popping it could be that there is debris that is partially clogging the probe. A convenient way to prime and clear debris is to seal the top of the LVC probe with a gloved finger, forcing solvent flow to go through the capillary. Several solvent flow compositions can be used. We have used combinations of methanol, chloroform, acetonitrile, water, and formic acid with success. The solvent flow rate will have to be tuned for each solvent composition in order to vortex. Typically, solvent flow rates <100 μL/min should be avoided as this results in wider and less intense signal profiles. It is key to keep the solvent surface tension low to enable vortexing, e.g., 100% water is difficult to vortex successfully.
9. The solvent composition will directly influence what molecules are extracted from the cells and are analyzed. Tuning the solvent composition for molecules of interest can significantly influence the sensitivity of the technique.
10. Here we use electrospray ionization as the ionization source of the system. Atmospheric pressure chemical ionization or essentially any other liquid-based ionization method can be used with the LVC probe.
11. If TIC background signal is highly variable it likely means that not enough solvent is being provided. If ablation signals are variable, adjust the heat shrink tubing to be closer to the sample slide. If background solvent ion signals are present but there is no signal from the dye then it likely means that too much solvent is being supplied and the LVC probe is partially overflowing.
12. When the LVC is moved into position to sample it will block light to the microscope. It can be helpful to mark where the cell of interest is using a marker in the LMD software. Alternatively,

fluorescence light (which comes through the top objective) can be used to visualize the sample while the LVC is in position to sample.

13. Spacing of the laser between raster lanes should be set to be slightly less than the diameter of the laser spot. This ensures that all sample material is ablated.

14. Selecting capture cap 'D' in the 'Collecting Tube Cap' prompt will cause the LVC-MS probe to collect signal when the laser ablates the outline of the shape. To avoid mixing laser ablation and 'Cut and Drop' signals, delay ejection of the microdissection until the laser ablation signal is eluted or select a different capture cap position in Collecting Tube Cap' prompt. Note that this requires that the section remains statically attached to the glass backing of the PEN membrane slide such that the section can be ejected at a later time.

15. Successful microdissection depends on a number of sample-specific factors. At times microdissection will seem to 'miss' the LVC probe. This is usually due to the section being ejected at extremely sharp angles, redeposition of the section onto the PEN membrane by static forces or the microdissection still being connected to the PEN membrane slide. There are several 'tricks' that can be used to increase the odds of successful microdissection.
    (a) It is critical to completely detach the microdissection from the PEN membrane slide but still leave enough surface area to remain attached to the glass backing through static attraction. It can help to cut the outline of the section using a small laser spot and then use a larger laser spot focused at the center of the section to eject the section.
    (b) Several repeats (2–3 times) of the 'Draw and Scan' cuttings can help to make a clean cut without ejecting the section. Performing several cuts at low laser power to cut the section and then one at high laser power to eject the section also works well.
    (c) Sometimes leaving a small connective piece to the section helps for it to remain attached to the PEN membrane slide. To eject the section simply cut the connective piece.
    (d) Defocusing the microscope will increase the laser spot diameter. This can help to successfully eject the section.

Acknowledgements

This work and J. F. C. was supported by the U.S. Department of Energy, Office of Science, Basic Energy Sciences, Chemical Sciences, Geosciences, and Biosciences Division and the

Laboratory Directed Research and Development Program of Oak Ridge National Laboratory, managed by UT-Battelle, LLC, for the U.S. Department of Energy. V. K. was supported by the U. S. Department of Energy, Office of Science, Basic Energy Sciences, Chemical Sciences, Geosciences, and Biosciences Division.

References

1. Zenobi R (2013) Single-cell metabolomics: analytical and biological perspectives. Science 342(6163):1243259. https://doi.org/10.1126/science.1243259
2. Wang D, Bodovitz S (2010) Single cell analysis: the new frontier in 'omics'. Trends Biotechnol 28(6):281–290. https://doi.org/10.1016/j.tibtech.2010.03.002
3. Pan N, Rao W, Kothapalli NR, Liu R, Burgett AWG, Yang Z (2014) The single-probe: a miniaturized multifunctional device for single cell mass spectrometry analysis. Anal Chem 86 (19):9376–9380. https://doi.org/10.1021/ac5029038
4. Shrestha B, Vertes A (2009) In situ metabolic profiling of single cells by laser ablation electrospray ionization mass spectrometry. Anal Chem 81(20):8265–8271. https://doi.org/10.1021/ac901525g
5. Cahill JF, Kertesz V (2018) Automated optically guided system for chemical analysis of single plant and algae cells using laser microdissection/liquid vortex capture/mass spectrometry. Front Plant Sci 9:1211. https://doi.org/10.3389/Fpls.2018.01211
6. Cahill JF, Kertesz V, Van Berkel GJ (2016) Laser dissection sampling modes for direct mass spectral analysis. Rapid Commun Mass Spectrom 30(5):611–619. https://doi.org/10.1002/rcm.7477
7. Shrestha B, Vertes A (2010) Direct analysis of single cells by mass spectrometry at atmospheric pressure. J Vis Exp (43):e2144. https://doi.org/10.3791/2144
8. Cahill JF, Kertesz V, Van Berkel GJ (2015) Characterization and application of a hybrid optical microscopy/laser ablation liquid vortex capture/electrospray ionization system for mass spectrometry imaging with sub-micrometer spatial resolution. Anal Chem 87(21):11113–11121. https://doi.org/10.1021/acs.analchem.5b03293
9. Cahill JF, Kertesz V, Porta T, LeBlanc JCY, Heeren RMA, Van Berkel GJ (2018) Solvent effects on differentiation of mouse brain tissue using laser microdissection 'cut and drop' sampling with direct mass spectral analysis. Rapid Commun Mass Spectrom 32(5):414–422. https://doi.org/10.1002/rcm.8053
10. Cahill JF, Kertesz V, Weiskittel TM, Vavrek M, Freddo C, Van Berkel GJ (2016) Online, absolute quantitation of propranolol from spatially distinct 20-and 40-μm dissections of brain, liver, and kidney thin tissue sections by laser microdissection-liquid vortex capture-mass spectrometry. Anal Chem 88(11):6026–6034. https://doi.org/10.1021/acs.analchem.6b01155

Chapter 8

Single Cell Analysis by High-Resolution Atmospheric-Pressure MALDI MS Imaging

Dhaka Ram Bhandari, Giulia Coliva, Maria Fedorova, and Bernhard Spengler

Abstract

Complementary to other imaging techniques, MALDI MS imaging provides label-free analysis and chemical specificity, covering a wide range of molecules that can be visualized simultaneously. Here, we highlight a protocol to visualize lipids from single cardiac cells using high-resolution atmospheric-pressure matrix-assisted laser desorption/ionization coupled to an orbital trapping mass analyzer.

Key words High-resolution atmospheric-pressure mass spectrometry imaging, Cardiac cells, Single cell imaging

1 Introduction

Mass spectrometry imaging (MSI) is a label-free analytical technique that is used to generate molecular images that represent distributions of analytes in cells, tissues, organs, or organisms. MSI is a robust tool capable of direct analysis of biomolecules from the surface of biological samples. Such direct analysis of biomolecules from tissue sections and single cells is vital for understanding biochemical processes of life [1]. Modern matrix-assisted laser desorption ionization (MALDI) MSI has shown high sensitivity and a high spatial resolution down to 1 μm [2, 3]. Using this technique, various classes of compounds [4] including lipids, peptides, proteins, and metabolites [5] have been visualized. The capability to focus the laser down to the subcellular level provides the possibility to better understand the functioning of cells, tissues, organs, or organisms. The technique has overcome the limitations of histochemistry and immunohistology in terms of specificity and the ability to detect multiple biomolecules simultaneously.

To understand the physiological processes and pathological stages in biomedical experiments cell cultures are widely used.

Bindesh Shrestha (ed.), *Single Cell Metabolism: Methods and Protocols*, Methods in Molecular Biology, vol. 2064, https://doi.org/10.1007/978-1-4939-9831-9_8,

The study of metabolites and visualizing them in a single cell provides essential information on phenotypic characteristics [6]. Lipid compositions and regulation in cell culture models are widely studied by high-resolution mass spectrometers, which, combined with chromatographic techniques, allows acquiring comprehensive information about the cellular lipidome. Nevertheless, no information about lipid localization can be obtained by LC-MS/MS analysis.

The study of subcellular lipid distribution is fundamental for understanding their role in cellular architecture and metabolic and signaling pathways. Conventional imaging techniques such as fluorescence and electron microscopy are widely used for lipid imaging on the tissue and subcellular level. However, they require target-specific labeling. Despite the high spatial resolution, microscopy imaging techniques are limited in multiplexing and do not provide identification of targeted analytes further than the presence of the labeled moiety. In contrast, MALDI MSI provides the possibility to perform label-free simultaneous imaging of hundreds of lipid species with minimal sample preparation. The identification of lipid species is supported by high-resolution, and high-mass-accuracy MS analysis and can be further improved by application of tandem MS.

Other mass spectrometric imaging techniques are available, such as desorption electrospray ionization (DESI) and secondary ion mass spectrometry (SIMS). DESI requires almost no sample preparation and analysis is performed at atmospheric pressure. However, its application to subcellular imaging is not possible since the spatial resolution achieved is limited to 30–50 μm by the solvent spray geometry. SIMS is an imaging technique that achieves the best values in terms of spatial resolution, making it the most used technique to visualize subcellular distributions. However, sample preparation requires careful optimization for sample compatibility to the ultrahigh-vacuum conditions in which the analysis is performed. Furthermore, severe prompt fragmentation of analyte ions complicates analysis and limits the applicability of SIMS, especially in the higher mass range.

Here, we present a protocol for application of a high-resolution MALDI imaging system (AP-SMALDI5 AF, TransMIT GmbH, Giessen, Germany) operating at atmospheric pressure, which provides easy sample handling and is compatible with a wide variety of MALDI matrices. Additionally, the system is capable of pixel-wise autofocusing for high-resolution surface analysis of nonplanar objects, for example, to visualize multiple lipid species in single primary rat cardiac cells, after fast sample preparation. A summarized workflow of single cell MALDI MSI experiment is shown in Fig. 1.

The protocol describes details on culturing cells on a glass coverslip that is compatible with high-resolution MALDI MS

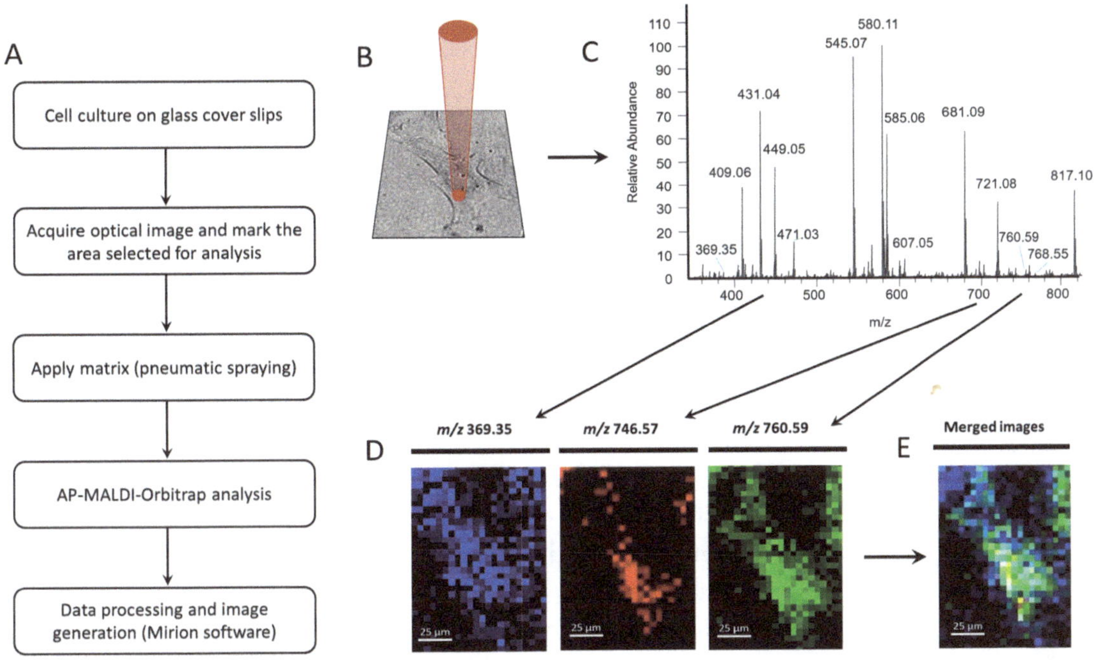

Fig. 1 Summarized workflow of single cell MALDI MSI experiments. (**a**) Main experimental steps involved in single cell MALDI MSI; (**b**) Illustration of data acquisition and analysis strategy illustrated by single-cell images of primary rat cardiac cells (**c**) A single MS spectrum (**d**) MS images for cholesterol, $[M+H-H_2O]^+$, *m/z* 369.35 blue, phosphatidylethanolamine (PE(36:1)) $[M + H]^+$, *m/z* 746.57, red and phosphatidylcholine (PC(34:1)) $[M + H]^+$ *m/z* 760.59, green, and (**e**) RGB overlay of the images. Pixel size = 5 μm and image bin width = ± 5 ppm

imaging, selection of cells for MALDI MS imaging experiments, matrix application, and MS data acquisition using the autofocusing feature of the MALDI MS imaging source with a laser focus of 5 μm. The high spatial resolution down to the micrometer range assists in visualizing metabolites from compartments of micrometer size. In order to resolve the chemical complexity, the MS spectra are generated with a high-resolution high-accuracy mass spectrometer operating at a mass resolution of 240,000 at *m/z* 200. MS ion images with a mass bin width of ±5 ppm are generated, and annotation of the signals is performed using online search engines such as human metabolome database (HMBD) [7], LIPID MAPS [8], Metlin [9], and ALEX [10].

2 Materials

2.1 Cell Culture

1. Primary rat cardiac cells (Innoprot, Elexalde Derio, Spain).
2. Attachment Factor (Life Technology GmbH, Darmstadt, Germany).

3. DMEM/F12 Dulbecco's Modified Eagle Medium/Ham's F-12 (Life Technology GmbH, Darmstadt, Germany).
4. Fetal bovine serum (Life Technology GmbH, Darmstadt, Germany).
5. Horse serum (Life Technology GmbH, Darmstadt, Germany).
6. Nonessential amino acids (Life Technology GmbH, Darmstadt, Germany).
7. Sodium pyruvate (Life Technology GmbH, Darmstadt, Germany).
8. L-glutamine (Life Technology GmbH, Darmstadt, Germany).
9. Penicillin-streptomycin (Life Technology GmbH, Darmstadt, Germany).
10. Phosphate buffered saline (Life Technology GmbH, Darmstadt, Germany).
11. Paraformaldehyde (Sigma Aldrich GmbH, Taufkirchen, Germany).
12. Ethanol (Carl Roth GmbH + Co. KG, Lörrach, Germany).
13. Borosilicate round glass coverslips, 13 mm diameter (VWR International GmbH, Dresden, Germany).
14. Water (HPLC grade, Honeywell International Inc., Seelze, Germany).

2.2 Sample Preparation for MALDI MS Imaging

1. Keyence VHX-5000 digital microscope (KEYENCE Deutschland GmbH, Neu-Isenburg, Germany).
2. Double-sided microscope sticker (PLANO GmbH, Wetzlar, Germany).
3. MALDI matrix, 30 mg/mL 2,5-dihydroxybenzoic acid (DHB, Merck Schuchardt OHG, Hohenbrunn, Germany).
4. Acetone (HPLC grade, Merck KGaA, Darmstadt, Germany).
5. Water (HPLC grade, Honeywell International Inc., Seelze, Germany).
6. Trifluoroacetic acid (HPLC grade, AppliChem GmbH, Darmstadt, Germany).
7. Pneumatic sprayer (SMALDIPrep, TransMIT GmbH, Giessen, Germany).

2.3 Instrumentation and Data Analysis for MALDI MS Imaging

1. AP-SMALDI5 AF MSI source (TransMIT GmbH, Giessen, Germany).
2. Orbital trapping mass analyzer Q Exactive family mass spectrometer (Thermo Scientific, Bremen, Germany).
3. MIRION software package (TransMIT GmbH, Giessen, Germany).

3 Methods

3.1 Preparation of Cells for MS Imaging

1. Sterile borosilicate glass coverslips were placed in a 24-well plate and covered with 0.5 mL of Attachment Factor (gelatin) for 1 h at 37 °C.
2. DMEM/F12 was supplemented with FBS (20%, v/v), 5% horse serum, 2 mmol/L L-glutamine, 3 mmol/L sodium pyruvate, 0.1 mmol/L nonessential amino acids, 1% penicillin-streptomycin.
3. Primary rat cardiac cells were seeded on the coverslips in a density of 1×10^4 mL^{-1} and let grow until 70% confluence (*see* **Note 1**).

   Optimal cell confluence for MALDI MSI experiments needs to be evaluated based on the cell type. To be able to study the subcellular lipid distribution, it is necessary to specifically target cells that are not overlapping or too closely located to the others.
4. The supplemented medium was removed, cells were washed with PBS, and incubated with DMEM/F12 without supplements for 5 h.
5. Cells were washed with PBS and fixed with PFA (4%, v/w; 37 °C) for 15 min.
6. Cells were washed once with PBS and twice with water.
7. Cells were stored in water at 4 °C before the analysis.

3.2 Sample Fixation on MALDI Target Holder

1. Selection of the cells for MALDI MS imaging was performed by using a Keyence VHX-5000 digital microscope.
2. To select acquisition area for MALDI MSI experiments, the coverslip was examined under the optical microscope, and an area with cells with optimal confluence and distribution was marked with a marker pen on the back side of the coverslip. Optical image of the coverslip was acquired before (to be used for colocalization with MALDI images) and after marking (to be used to identify the area selected for analysis when the coverslip is loaded into the MALDI source).
3. The coverslip with the cell culture was fixed on the MALDI target plate using a double-sided microscope sticker.

3.3 Matrix Application

1. The sample on the MALDI target plate was kept in the sample holder of the pneumatic sprayer (SMALDIPrep, TransMIT GmbH, Giessen, Germany).
2. DHB was used as a MALDI matrix (30 mg/mL; in acetone/water (v/v) containing 0.1% TFA) as the peak intensity was highest for lipid analysis from the tissue surfaces.

3. The MALDI matrix solution was sprayed using the pneumatic sprayer system ($V = 100$ μL).
4. Nitrogen was used as the carrier gas at a pressure of 1 bar.
5. The flow rate was maintained at 10 μL/min using a syringe pump.
6. The sample was rotated at a speed of 350 rpm.
7. Crystal size, matrix layer quality, and homogenous coverage were checked with an optical microscope. These parameters are crucial for high-resolution MS imaging, as they directly influence the generated MS images. If these parameters are optimized for MS imaging, normalization, and processing of MS images are not required.

3.4 Data Generation

1. After the matrix was applied to the sample, the MALDI target was mounted to the sample holder of the MALDI imaging source.
2. The area to be scanned by the laser was selected using the observation camera that displays the surface of the sample.
3. To focus the laser down to 5 μm, the autofocus feature of the source was used. Using the autofocus feature, the laser focus is precisely lowered down to the required size. The focusing of the laser is performed in less than a second (*see* **Note 2**).
4. A low laser pulse energy was selected that permits to measure with the required lateral resolution. This is achieved by choosing the maximum intensity of the peaks without oversampling the region of interest; the control software of AP-SMALDI5 AF was used to select the lateral resolution (step size, e.g., 5 μm) and the number of pixels (*see* **Note 3**).
5. For each pixel, ions from 50 laser pulses were collected in the c-trap of the mass spectrometer before transferring them to the Orbital trapping mass analyzer. This number of laser pulses is required for the optimal intensity of the MS peaks.
6. The samples were scanned with 5 μm step size as this provided the optimal intensity for low abundance compounds. Lowering the step size further, reduced the peak intensity and led to oversampling.
7. Highly resolved mass spectra were generated with a Fourier transform orbital trapping mass spectrometer (Exactive or Q Exactive, Thermo Fisher Scientific GmbH, Bremen, Germany) (*see* **Note 4**).
8. The target voltage was set to 3 kV.
9. The mass spectrometer was operated in the positive-ion mode.
10. Spectra were scanned in the mass-to-charge-number (m/z) range of 300–1200.

11. The mass resolution (R) was set at 240,000 at *m/z* 200 (*see* **Note 5**).
12. Automatic gain control was disabled and the ion injection time was set to 500 ms.
13. Internal calibration was achieved by using lock masses from a matrix cluster ion $[5\ DHB\text{-}4\ H_2O + NH_4]^+$ at *m/z* 716.12461.
14. The cycle time in pixel mode was 1 scan/s.

3.5 Image Generation

1. MSI images were obtained using the MIRION software package [11] using the *.udp file generated by the AP-SMALDI5 AF control software.
2. MIRION reads the *.raw file using the *.udp file.
3. For targeted lipid visualization, MSI images of selected lipid species were generated by using exact theoretical mass values with an *m/z* bin width of ±5 ppm. Due to the relatively low signal intensities usually obtained in single cell MALDI MS imaging experiments, acquisition of additional tandem MS data for structural identification of assigned species was not possible. Thus, identification of lipids relies solely on accurate mass; therefore, high mass resolution and high mass accuracy are essential. It is beneficial to perform a complementary experiment using LC-MS/MS based lipidomics on the same sample set, e.g., identical cell culture experiment followed by a pellet collection, lipid extraction, and LC-ESI-MS/MS analysis using data-dependent acquisition. Lists of lipids identified in the LC-MS/MS experiments can be further used for the assignment of lipid species in MALDI MSI datasets. However, it is important to keep in mind the difference between MALDI and ESI ionization, especially in terms of adduct formation (e.g., protonated vs. sodium and potassium adducts). Additionally, matrix adducts (136.01 u mass increment in case of DHB matrix) can be observed in MALDI MSI datasets.
4. For untargeted lipid visualization, due to the high number of peaks generated in the experiments, additional filtering can be applied using the MIRION software. The software provides the possibility of adding several filters to reduce the number of images generated from chemical noise. For instance, it is possible to observe images only within a specific *m/z* range, or defined mass defect range to exclude DHB matrix cluster peaks or to determine the number of pixels covered in the images. If the number of covered pixels in an image is too high, it means that the signals were acquired from the blank area surrounding the cells. A signal detected from almost every pixel might correspond to a matrix compound. On the other hand, a lower threshold of pixel coverage is used to exclude noise signals. The best range of pixel coverage to be analyzed is

specific for each sample as it depends on the cell confluence of the area to be analyzed.

5. The selected ion images were normalized to the highest intensity for each ion species separately.
6. Red-green-blue (RGB) images of three different *m/z* values were created using MIRION.
7. Images were exported with their corresponding centroided (intensity weighted) *m/z* values as ∗.jpg or ∗.bmp.
8. Other data processing steps like interpolation, smoothing, or normalization to matrix signals were not required and thus, were not applied during the image generation process.

3.6 Annotation of the Images

1. Identified compounds were assigned using high mass accuracy (<1 ppm).
2. Databases such as METLIN, HMBD, LIPID MAPS, and ALEX lipid calculator were used for lipid annotation.
3. Mass accuracies were calculated using intensity-weighted average values from all spectra of one image.

4 Notes

1. Samples should be handled properly to avoid contamination.
2. For transparent and non-scattering samples, the autofocus operation does not work properly as the autofocus depends on the reflection of light from the surface.
3. Lateral resolution (step size) and the number of pixels should be selected in such a way as to limit the needed experimental time.
4. Higher mass resolution and a higher number of pixels increase the measurement time.
5. High mass resolution is important when signal intensities are too low for tandem mass spectrometry.

Acknowledgments

Financial support by the Deutsche Forschungsgemeinschaft (DFG) under grant Sp314/13-1. M.F. and G.C. are financially supported by funding from the European Commission's Horizon 2020 research and innovation programme under the Marie Sklodowska-Curie grant agreement number 675132 (MSCA-ITN-ETN MASSTRPLAN). Financial support from the German Federal Ministry of Education and Research (BMBF) within the framework of the e:Med research and funding concept for SysMedOS project (to MF) is gratefully acknowledged.

Competing financial interests: B.S. is a consultant to TransMIT GmbH Giessen. D.B. is a part-time employee of TransMIT GmbH.

References

1. Spengler B (2015) Mass spectrometry imaging of biomolecular information. Anal Chem 87 (1):64–82
2. Kompauer M, Heiles S, Spengler B (2016) Atmospheric pressure MALDI mass spectrometry imaging of tissues and cells at 1.4-μm lateral resolution. Nat Methods 14:90
3. Zavalin A, Yang J, Hayden K, Vestal M, Caprioli RM (2015) Tissue protein imaging at 1 μm laser spot diameter for high spatial resolution and high imaging speed using transmission geometry MALDI TOF MS. Anal Bioanal Chem 407(8):2337–2342
4. Römpp A, Spengler B (2013) Mass spectrometry imaging with high resolution in mass and space. Histochem Cell Biol 139(6):759–783
5. Bhandari DR, Schott M, Römpp A, Vilcinskas A, Spengler B (2015) Metabolite localization by atmospheric pressure high-resolution scanning microprobe matrix-assisted laser desorption/ionization mass spectrometry imaging in whole-body sections and individual organs of the rove beetle Paederus riparius. Anal Bioanal Chem 407 (8):2189–2201
6. Svatoš A (2011) Single-cell metabolomics comes of age: new developments in mass spectrometry profiling and imaging. Anal Chem 83 (13):5037–5044
7. Wishart DS, Jewison T, Guo AC, Wilson M, Knox C, Liu Y, Djoumbou Y, Mandal R, Aziat F, Dong E, Bouatra S, Sinelnikov I, Arndt D, Xia J, Liu P, Yallou F, Bjorndahl T, Perez-Pineiro R, Eisner R, Allen F, Neveu V, Greiner R, Scalbert A (2013) HMDB 3.0—The Human Metabolome Database in 2013. Nucleic Acids Res 41(D1):D801–D807
8. Fahy E, Sud M, Cotter D, Subramaniam S (2007) LIPID MAPS online tools for lipid research. Nucleic Acids Res 35(suppl_2): W606–W612
9. Smith CA, O'Maille G, Want EJ, Qin C, Trauger SA, Brandon TR, Custodio DE, Abagyan R, Siuzdak G (2005) METLIN: a metabolite mass spectral database. Ther Drug Monit 27(6):747–751
10. Husen P, Tarasov K, Katafiasz M, Sokol E, Vogt J, Baumgart J, Nitsch R, Ekroos K, Ejsing CS (2013) Analysis of lipid experiments (ALEX): a software framework for analysis of high-resolution shotgun lipidomics data. PLoS One 8(11):e79736
11. Paschke C, Leisner A, Hester A, Maass K, Guenther S, Bouschen W, Spengler B (2013) Mirion—a software package for automatic processing of mass spectrometric images. J Am Soc Mass Spectrom 24(8):1296–1306

Chapter 9

A MALDI-MS Methodology for Studying Metabolic Heterogeneity of Single Cells in a Population

Jasmin Krismer, Jens Sobek, Robert F. Steinhoff, Rolf Brönnimann, Martin Pabst, and Renato Zenobi

Abstract

Mass spectrometry based metabolomics is the highly multiplexed, label-free analysis of small molecules such as metabolites or lipids in biological systems, and thus one of the most direct ways to characterize phenotypes. However, the phenotyping of populations with single-cell resolution is a great challenge due to the small number of molecules contained in an individual cell. Here we describe a microarray-based sample preparation workflow for MALDI mass spectrometry that has single-cell sensitivity and allows high-throughput analysis of lipids and pigments in single algae cells. The microarray targets receive individual cells in 1430 separate spots that allow the cells to be lysed individually without cross-contamination. Using positive ion mode and 2,5-dihydroxybenzoic acid as the MALDI matrix, the mass spectra unveil information about the relative composition of more than 20 different lipids/pigments in each individual cell within the population. Thus, the method allows the analysis of cellular phenotypes in a population on a completely new level.

Key words Single-cell analysis, MALDI-mass spectrometry, High-throughput analysis, *Chlamydomonas reinhardtii*, Lipid profiling

1 Introduction

Mass spectrometry (MS) is a pivotal tool for the analysis of metabolites in biological systems. Depending on the efforts taken, MS-based approaches can provide high sample throughput, highly multiplexed readouts, and/or exact quantitation of metabolites. The main requirements for the development of a successful single-cell mass spectrometry method are high throughput and high sensitivity and selectivity with respect to a very complex biological matrix [1].

The main ionization methods employed in the analysis of intact biomolecules are electrospray ionization (ESI) and matrix-assisted laser desorption/ionization (MALDI). MALDI is a soft ionization technique [2, 3] that allows the ionization and detection of intact

Bindesh Shrestha (ed.), *Single Cell Metabolism: Methods and Protocols*, Methods in Molecular Biology, vol. 2064, https://doi.org/10.1007/978-1-4939-9831-9_9,

biomolecules. It relies on the absorption of laser light and the transfer of charges by auxiliary matrix molecules. This ionization process, which is still not fully understood, results almost exclusively in the formation of singly charged ions greatly, which facilitates spectral interpretation [4, 5]. Most commercial instruments employ a UV-laser and a time-of-flight mass analyzer and feature very short spectral acquisitions times. Thus, they permit collecting of spectra in high-throughput fashion and allow spectral mapping of tissues or even whole organisms with high spatial resolution [6]. In selected examples, specialized laser optics and careful sample preparation allow the imaging of tissue slices with single-cell resolution [7]. Thus, MALDI is sensitive enough to provide spectra collected from the analytes contained in a single cell [8].

However, the analysis of single cells from liquid cell cultures by MALDI-MS struggles with reproducibility in sample preparation and the lysis of microbial cells. This is why most attempts at using MALDI for analyzing single cells show very limited throughput.

The workflow described here employs instrumentation that is widely used in microarraying technology, namely, a noncontact piezo-actuator-driven sub-nanoliter dispenser [9]. This technology imparts high reproducibility and accuracy to all the steps of sample preparation. Furthermore, we combined this with a dedicated MALDI target slide that provides 1430 distinct spots where individual cells can be placed for further processing and extraction (*see* Fig. 1). First the spots are covered by a layer of MALDI matrix (DHB). Then the cells are applied to the spots on the target by spotting a cell suspension featuring an optimized cell concentration aimed at yielding the maximal number of single-cell events. Since the cells emit fluorescence upon irradiation with specific laser wavelengths, their fluorescence can easily be detected on the metal MALDI target slides. Then their metabolism is quenched and mechanical damage inflicted by fast-freezing the cells in liquid nitrogen. Only then the matrix dissolved in organic solvents is applied to extract lipophilic analytes and co-crystallize them with co-precipitating MALDI matrix during evaporation of the solvents.

In this fashion, this protocol for MALDI-MS can avoid the heavy dilution that usually hampers ESI analyses. The protocol was optimized using *Chlamydomonas reinhardtii* populations grown in liquid cultures. *C. reinhardtii* is a single-celled eukaryote and a regularly used model organism for the study of photosynthesis [10] and more recently for algal biofuel production [11]. Its highly robust glycoprotein cell wall is a good surrogate for microbial systems [12]. Therefore, the protocol can potentially be adapted to other microbial systems that show auto-fluorescence (*see* Fig. 2). Using 2,5-dihydroxybenzoic-acid as a matrix [13] in the positive ion mode, the method gives access to approximately 20 different lipids and pigments in *C. reinhardtii* (*see* Table 1). The identity of compounds was confirmed using FT-ICR mass spectrometry (*see*

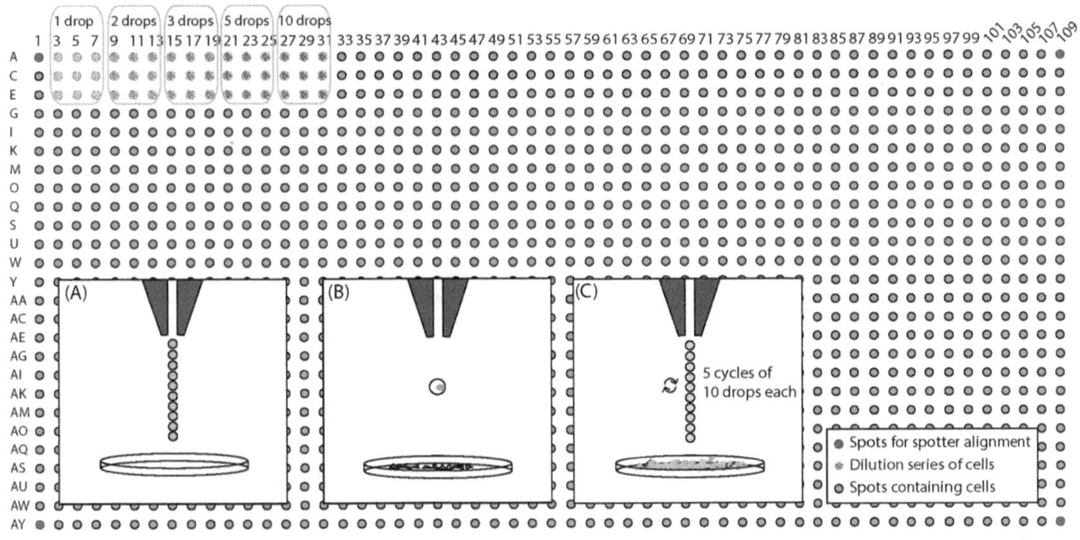

Fig. 1 Schematic of the array. Spots on the edge (A1, A109, AY1, AY109) can optionally serve for spotter alignment. (**a**) All the wells are covered with a layer of matrix (10 drops of the DHB solution). (**b**) After the optimal number of drops is defined, the whole array is covered with cell suspension resulting in many single-cell-covered wells. (**c**) Analytes are extracted from cells by recurrent application of MALDI matrix dissolved in acetone/water

Fig. 3) taking advantage of the very high mass accuracy and of CID based MS/MS measurements that were carried out at the population level. From the 1430 spots on the target slide, ideally around two-thirds would be occupied by single cells. Thus, one measurement can yield around 1000 single-cell spectra per chip. The highly reproducible spectra are a highly interesting target for statistical analyses with respect to the detected signals as well as their relative abundance. This allows for categorization and characterization of different phenotypes within the population.

2 Materials

2.1 MALDI Target Slides

2.1.1 Chemicals/Materials

1. Stainless steel plate (X5CrNi18 10, 1.25 mm thickness).
2. Acetone, HPLC quality (Sigma Aldrich, Chromasolv).
3. Sonication bath (Bandelin, Sonorex).
4. Non-fuzzing wipes.
5. Glass jar for sonication.
6. Water, for chromatography (Merck, LiChromasolv).
7. Chloroform (Sigma Aldrich, ReagentPlus).
8. Isopropanol, HPLC quality (Sigma-Aldrich, Chromasolv).
9. 50 ml plastic tubes with screw caps, polypropylene.

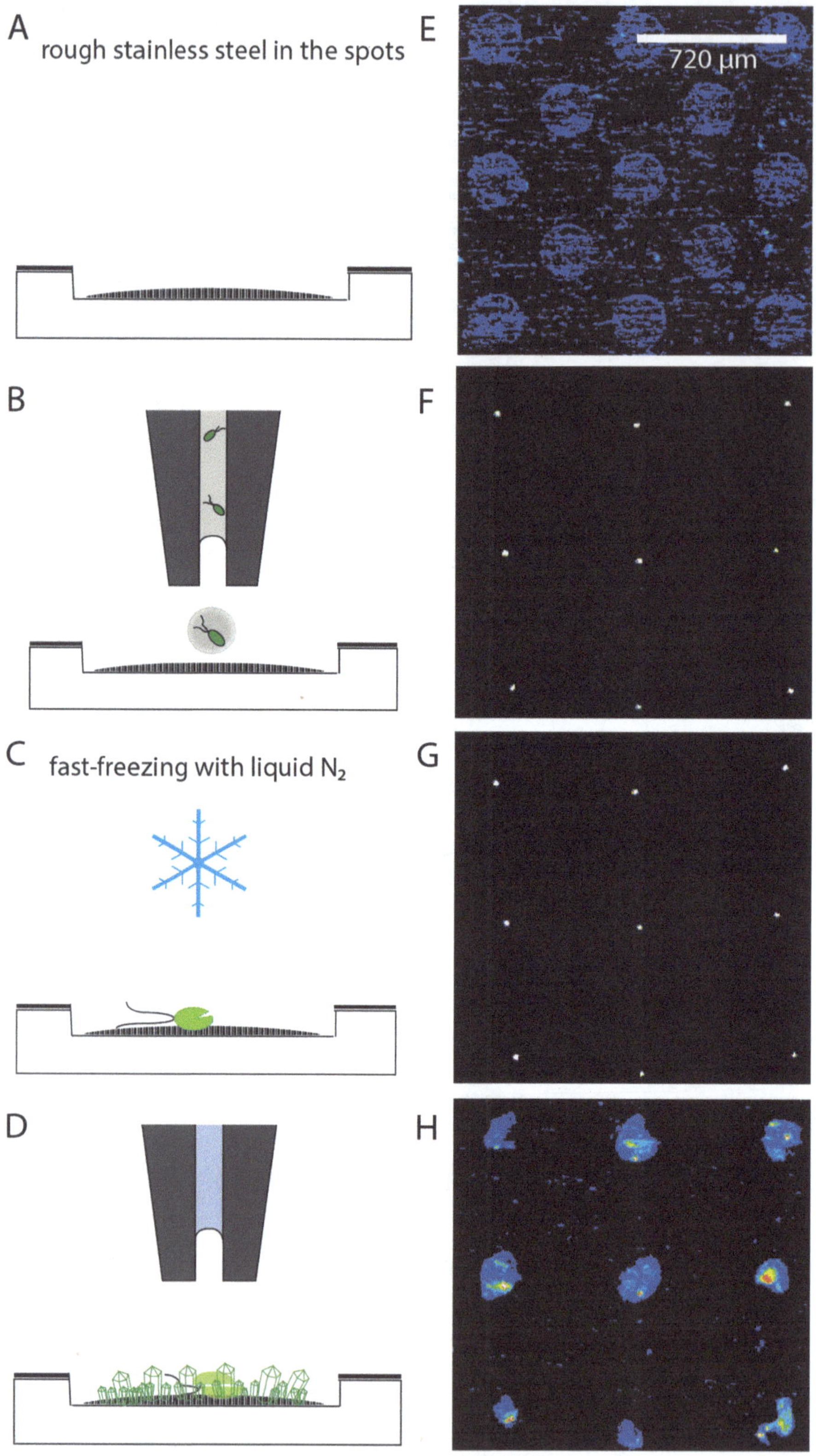

Fig. 2 Schematic of workflow and fluorescence scanning (ex. 630 nm, em. 670 nm). (**a/e)** shows the empty slide that is scanned to check for auto-fluorescence contamination on the chip. (**b/f**) After the cells are spotted they can be easily visualized on the slide. This is important to assign the mass spectra and to check for early

2.2 Single-Cell Sample Preparation

2.2.1 Chemicals/ Materials

1. Fluorescence dye 5 μM in 3xSSC buffer for spotter alignment (e.g., CY3).
2. MALDI-matrix solution (10 mg/ml 2,5-dihydroxybenzoic acid (Sigma Aldrich, 85707) in 90% acetone (Sigma Aldrich, Chromasolv) and 10% water (Merck, LiChromasolv), freshly prepared.
3. *Chlamydomonas reinhardtii* cells in the form of liquid suspension cell cultures (*see* **Note 1**).
4. 1.5 ml safe-lock, tapered-bottom vial, polypropylene.
5. 384-micro well tapered-bottom microplate, polypropylene.

Table 1
Compounds detected in single-cell FT-ICR-MS

	Compound name	Formula	Measured *m/z*	Theoretical *m/z*	δ (ppm)
Thylakoid lipids	MGDG (34:7) + Na+	$C_{43}H_{68}O_{10}$	767.47026	767.47047	0.3
	MGDG (34:7) + K+	$C_{43}H_{68}O_{10}$	783.44431	783.44441	0.1
	DGDG(34:6) + Na+	$C_{49}H_{80}O_{15}$	931.53888	931.53894	0.1
	DGDG(34:5) + Na+	$C_{49}H_{82}O_{15}$	933.55462	933.55459	0.0
	DGDG(34:3) + Na+	$C_{49}H_{86}O_{15}$	937.58611	937.58589	0.2
	DGDG(34:2) + Na+	$C_{49}H_{88}O_{15}$	939.60178	939.60154	0.2
Membrane lipids	DGTS (32:4)	$C_{42}H_{73}NO_7$	704.54577	704.54598	0.3
	DGTS (32:3)	$C_{42}H_{75}NO_7$	706.56126	706.56163	0.5
	DGTS (32:2)	$C_{42}H_{77}NO_7$	708.57761	708.57728	0.5
	DGTS (32:1)	$C_{42}H_{79}NO_7$	710.59311	710.59293	0.3
	DGTS (34:5)	$C_{44}H_{75}NO_7$	730.56125	730.56163	0.5
	DGTS (34:4)	$C_{44}H_{77}NO_7$	732.57698	732.57728	0.4
	DGTS (34:3)	$C_{44}H_{79}NO_7$	734.59267	734.59293	0.4
	DGTS (34:2)	$C_{44}H_{81}NO_7$	736.60863	736.60858	0.1
	DGTS (36:5)	$C_{46}H_{79}NO_7$	758.59257	758.59293	0.5
	DGTS (36:4)	$C_{46}H_{81}NO_7$	760.60822	760.60858	0.5
	DGTS (36:3)	$C_{46}H_{81}NO_7$	762.62399	762.62423	0.3
Pigments	Chlorophyll a (−Mg+3H+)	$C_{55}H_{74}N_4O_5$	871.57313	871.5732	0.1
	Chlorophyll b (−Mg+3H+)	$C_{55}H_{72}N_4O_6$	885.55288	885.55246	0.5
	Carotenoid (canthaxanthin)	$C_{10}H_{52}O_2$	565.40431	565.50501	0.5
	Carotenoid (fritschellaxanthin)	$C_{40}H_{54}O_3$	583.41505	583.41457	0.8
	Carotenoid (violaxanthin/ neoxanthin)	$C_{40}H_{56}O_4$	601.42546	601.42514	0.5

Fig. 2 (continued) lysis. (**c/g**) The cells are quenched and mechanically severed by fast freezing in liquid nitrogen. Cell fluorescence remains unchanged. (**d/h**) The analytes are extracted from the cells and co-crystallized by repeated spotting of DHB in 90% aqueous acetone (Copyright © American Society for Microbiology, Appl Environ Microbiol 81:5546–5551. https:/doi.org/10.1128/AEM.01201-15)

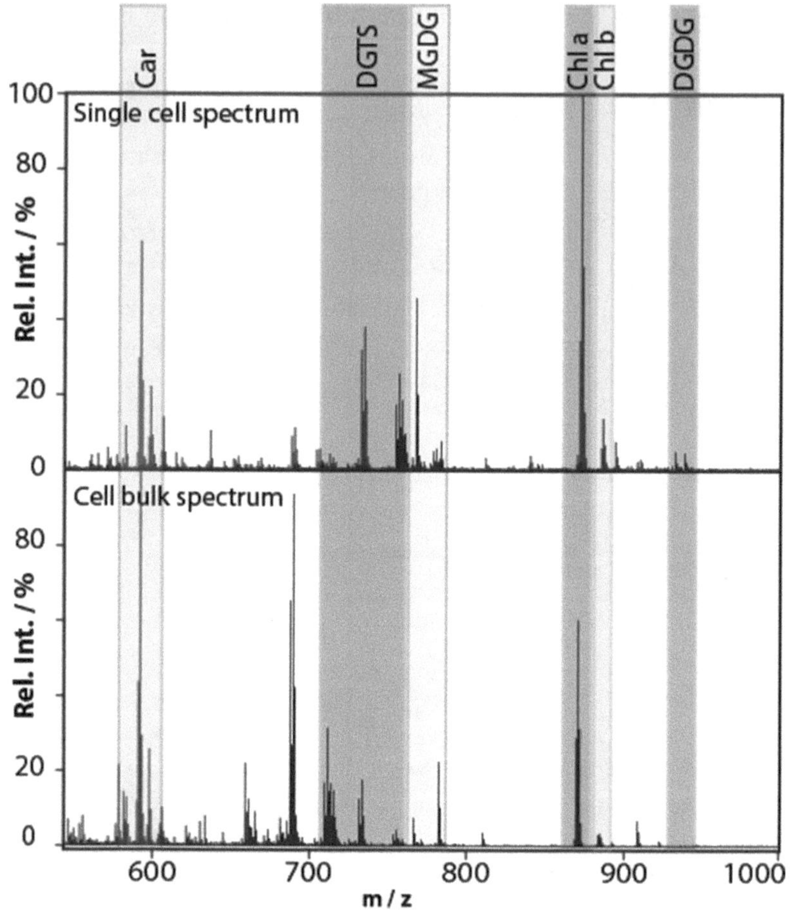

Fig. 3 MALDI mass spectra obtained from a single cell (upper spectrum) and from a bulk *C. reinhardtii* (lower spectrum) on an FT-MS instrument. The single-cell mass spectrum is background corrected for better visibility of the peaks in the lower mass region. This was achieved by subtracting the spectrum of an empty spot from the single-cell spectrum. The main analyte classes that are detected are carotenoids (Car), homoserine-lipids(DGTS), glycolipids (MGDG and DGDG), and chlorophylls (Chl a and Chl b) (Copyright © American Society for Microbiology, Appl Environ Microbiol 81:5546–5551. https:/doi.org/10.1128/AEM.01201-15)

6. Liquid nitrogen.
7. Long tweezers.

2.2.2 Instrumentation

1. Confocal fluorescence scanner (Tecan LS 400) (*see* **Note** 2).
2. Noncontact microarray spotter (noncontact piezo-actuator-driven spotting device, e.g., Scienion S11) (*see* **Note** 3).
3. Centrifuge for 1.5 ml tubes (5415R, Eppendorf).
4. Desiccator connected to a vacuum/nitrogen line.

2.3 MALDI-MS Analysis of Single Cells

2.3.1 Chemicals/Materials

1. Chlorophyll analytical standard (Sigma Aldrich, C5753).
2. Freshly prepared MALDI-matrix solution (10 mg/ml 2,5-dihydroxybenzoic acid (Sigma Aldrich, 85707) in 90% acetone (Sigma Aldrich, Chromasolv) and 10% water (Merck, LiChromasolv).

2.3.2 Instrumentation

1. Mass spectrometer equipped with a UV-MALDI source (*see* **Note 4**).

3 Methods

3.1 MALDI Target Slides

Manufacturing Target Slides

1. MALDI targets are made from stainless steel slides (75 mm × 25 mm) and cut with high precision (>100 μm) using laser from a big piece of stainless steel (30 × 30 cm) of 1.25 mm thickness using laser machining (Nd:YAG laser).
2. The slides are microstructured with the single-cell array spots using picosecond laser ablation. The array consists of 1430 spots (26 rows, 55 columns) with a diameter of 300 μm each and a 720 μm center-to center distance. The spots showed a rough surface and were less than 1 μm deep. They serve as marks for orientation during sample preparation and MALDI-MS measurements and their rough surface facilitates the "focusing" of liquids and the crystallization of the matrix.
3. Slides should always be cleaned properly before use (*see* **Note 5**).

Cleaning MALDI Target Slides

1. MALDI target slides are scrubbed with an acetone-drenched non-fuzzing wipe.
2. Slides are sonicated in a glass jar using solvents with different polarities. The following solvent conditions should be maintained for 30 min each in a sonication bath: 100% water, 100% isopropanol, 100% acetone, 100% chloroform.
3. After this cleaning procedure the slides are stored in 50 ml falcon tubes in a nitrogen atmosphere.

3.2 Single-Cell Sample Preparation

1. To check for potential contamination the stainless steel slides are scanned (*see* Fig. 2) for chlorophyll fluorescence before the sample application (excitation at 630 nm, detection at 670 nm, detector gain: 120.6 μm resolution). Scanning parameters are kept constant for the rest of the experiment.
2. The x–y stage of the pirzo dispenser is aligned with the prestructured array on the slide by spotting a fluorescent dye (e.g., CY3) into the corner spots on the slide (*see* Fig. 1). Adjust

positioning of the droplet spotter if necessary (*see* **Note** 6). Scan the alignment marks at 532 nm excitation and 570 nm detection.

3. A layer of matrix is spotted into each recipient spot by applying 10 drops of a 10 mg/ml solution of DHB dissolved in 90% aqueous acetone.
4. To minimize the ion suppressing effect of the salts contained in the culture media during MALDI-MS measurements, cells are centrifuged at 2000 × *g* and resuspended in deionized water. This procedure is repeated 3 times (*see* **Note** 7).
5. The cell number is adjusted to approximately 10^6 cells per ml (1 cell per nanoliter, 0.5 cells per drop) using deionized water.
6. To determine the optimal number of drops for highest yield of single-cell events on the chip a dilution series is spotted; 3 × 3 spots are filled with 1, 2, 3, 5, and 10 drops of the cell suspension.
7. The part of the slide that contains the dilution series is scanned and the number of drops that show the highest yield of single cells is determined (needed for **step 8**).
8. The cells are spotted using the previously defined number of drops into all the wells. A fluorescence scan is used to determine the number of cells in each spot. This information is essential to assign the mass spectra.
9. The slide is submerged in liquid nitrogen to quench the metabolism and to facilitate the analyte extraction (*see* **Note** 8).
10. The slide is returned to room temperature in a desiccator to prevent condensation.
11. The analytes are extracted by spotting the matrix dissolved in 90% aqueous acetone; 5 cycles of 10 drops are applied resulting in recurrent and more reproducible crystallization of the analytes and the matrix (*see* **Note** 9).
12. A fluorescence scan of the slide is performed after matrix application to check the lysis of the cells and the reproducibility of the extraction (*see* Fig. 4).
13. The slide is stored at −80 °C in a 50 ml falcon tube filled with nitrogen for MS measurements.

3.3 MALDI-MS

1. The target slides are reconstituted to room temperature just before the MALDI measurements in vacuum to avoid condensation of humidity.
2. The target slides are complemented with external chlorophyll standards premixed with MALDI matrix solution for external mass calibration.

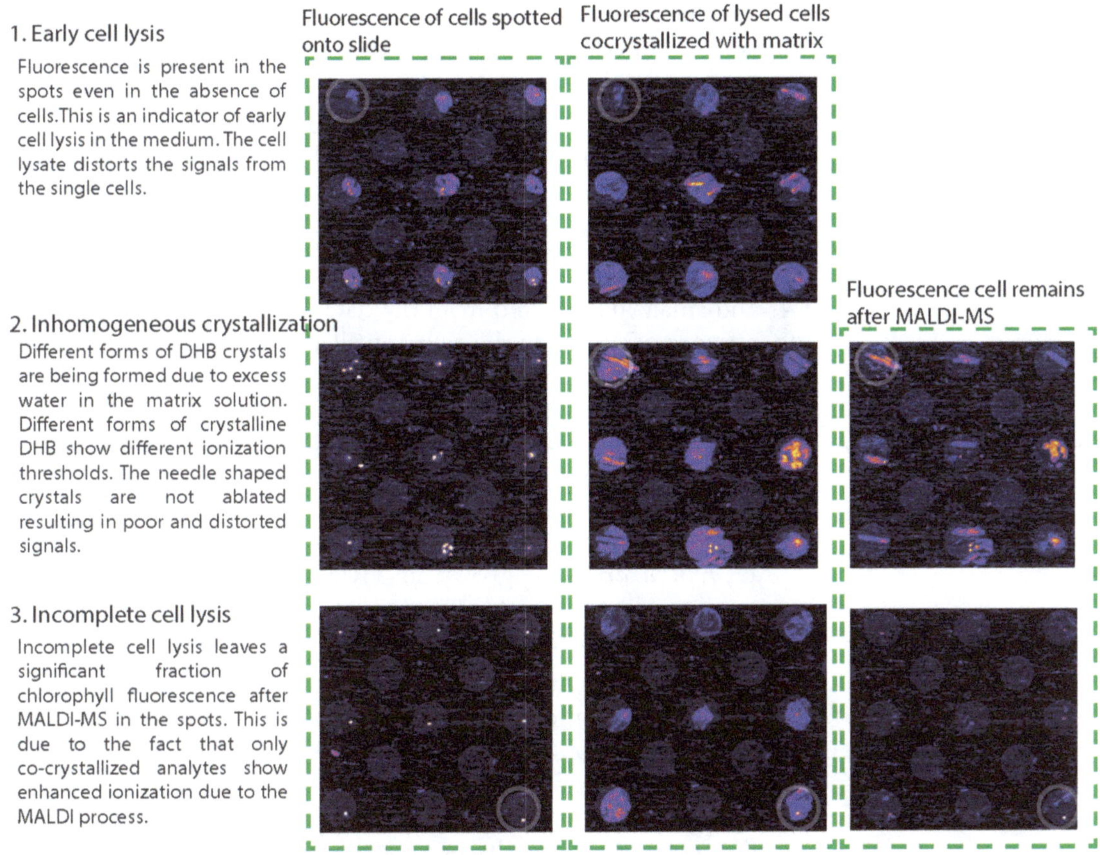

Fig. 4 Diagnosis for different problems that can occur during sample preparation process. Early cell lysis (1) leads to a corrupted readout that no longer reports single-cell properties. Inhomogeneous crystallization (2) and incomplete cell lysis (3) result in low signal response. In both cases, the majority of the analytes are not ionized in the first case because the crystals are not ablated in the latter because the analytes are not being extracted (Adapted from Krismer et al. 2015 Appl Environ Microbiol 81:5546–5551. https:/doi.org/10.1128/AEM.01201-15 Copyright © American Society for Microbiology)

3. The slides are loaded into the mass spectrometer and MALDI mass spectra are recorded in automated fashion for every spot using a spiral ablation pattern. The following conditions were successful on an AB Sciex 5800 MALDI TOF instrument: mass range 500–1000 m/z, 4300 a.u. laser energy, 400 Hz pulse rate, detector voltage: 1.770 kV, 200 ns delayed extraction time, 12 shots/sub-spectrum, 50 sub-spectra. Measurement time was 2 h 45 min per slide, 6.5 s per spot.

3.4 Data Analysis and Interpretation

1. The number of cells, as extracted from the fluorescence scan in **step 8** (*see* Subheading 3.2, **step 8**) should be attributed to each mass spectrum.
2. Peak lists containing the information on the m/z value, the peak area, the S/N value, and the resolution of each signal with S/N > 10 are exported for all of the measured spots.

3. To check for cross-contaminations and lysis efficiency the dataset should comply with the following criteria (for more information *see* **Notes 9–12**):
   (a) Mass spectra of spots that lack auto-fluorescence must not show cell-derived signals but show signals from of 2,5-DHB.
   (b) More than 90% of the spots occupied by a single cell should detect chlorophyll a.
4. Information extracted from the data includes the presence and absence of signals in the cells as well as the relative composition of the detected signal in the cells.

4 Notes

1. So far the protocol was used most successfully applied to *C. reinhardtii* cells grown in TRIS-acetate phosphate (TAP) medium. However, this protocol can potentially be optimized for any fluorescent microbe.
2. Slide scanning after cell spotting and lysis, respectively, is important to measure the number of cells present in each spot, and for optimizing the lysis protocol. Chlorophyll fluorescence, excited at 633 nm and detected at 670 nm, shows little background fluorescence by scattering or DHB autofluorescence. Thus, the presence of chlorophyll can be traced by fluorescence scanning at all stages of the experiment (including lysis and co-crystallization).
3. The spotter should offer highly reproducible, customizable, noncontact spotting of drops in the sub-nanoliter range. For our experiments we used an S11 spotter from Scienion AG, Berlin.
4. The most important requirement on the mass spectrometer side is the possibility to run automated measurements. This should include programming the geometry of the array, alignment of the stage, and the laser with the spots as well as a feature for automatic measurements. We found that all mass spectrometers tested so far including AB Sciex 4800, AB Sciex 5800, and a Bruker SolariX 9.4 T provide enough sensitivity to measure single-cell spectra. FT-MS measurements showed a decrease in signal caused by MALDI-matrix sublimation when run times exceeded 6 h.
5. Laser ablation can result in oxidation of the stainless steel surface. Incubating the chip in 15% (aqueous) acetic acid at 60 °C in a sonication bath dissolves the oxidized part of the surface of the slide.

6. Proper alignment of the droplet dispensing system with the prestructured array is essential since the MALDI-MS stage is aligned with the array geometry for MALDI-MS. Furthermore, the rough spots on the stainless steel can aid to some extent in the refocusing of drops for the matrix. Thus, we always check the alignment of a new series of slides spotting a 5 μM solution of a dye-labeled fluorescent oligonucleotides into spots on the edges of the array (*see* Fig. 1).
7. Centrifugation speed. We observed that altering the cell culture conditions by using different media resulted in variations in lysis efficiencies. Optimizing centrifugation conditions during washing the cells can mostly compensate these effects.
8. Fast freezing with liquid nitrogen is essential for two reasons: first, the fast freezing leads to quenching of the cellular metabolism. It is important to minimize the time between the substitution of the culture medium and the freezing. Second, the freezing step greatly enhances the lysis efficiency while preventing cross-contamination on the chip.
9. Long run times (more than 6 h) combined with MALDI matrices that undergo sublimation processes under the vacuum conditions found in the ion source can result in a decrease in sensitivity during a measurement. Thus, it is important to monitor the total ion counts of the analyzed spots, which should be at a constant average throughout the measurement.
10. The highly acidic conditions during crystallization of DHB convert the chlorophyll into pheophytins. Under the conditions applied, this process was found to be very fast since no intact chlorophyll can be detected by MALDI-MS or by HPLC-UV.
11. The MALDI-MS measurements should generate spectra representative of the respective cells. The laser ablation of the spotted cell material should be complete, such that in a subsequent MALDI experiment on the same spot no signals are detected. This can be validated by a fluorescence scan of the slide after the MALDI experiment that is expected to show no remaining fluorescence. If this is not the case, the cell lysis procedure should be further optimized (*see* Fig. 4).
12. Each cell can only be measured once due to the limited amount of analytes present in a cell. For quantitative analyses of the spectra, instrumental noise should be taken into consideration. We therefore recommend only to consider peaks with $S/N > 50$. Under these conditions we found the instrumental noise to contribute less than 25% as tested by evaluating the isotope ratio variations (data not shown).

References

1. Zenobi R (2013) Single-cell metabolomics: analytical and biological perspectives. Science 342(6163):1243259
2. Fenn JB et al (1989) Electrospray ionization for mass spectrometry of large biomolecules. Science 246(4926):64–71
3. Karas M et al (1987) Matrix-assisted ultraviolet laser desorption of non-volatile compounds. Int J Mass Spectrom Ion Process 78:53–68
4. Karas M, Glückmann M, Schäfer J (2000) Ionization in matrix-assisted laser desorption/ionization: singly charged molecular ions are the lucky survivors. J Mass Spectrom 35(1):1–12
5. Zenobi R, Knochenmuss R (1998) Ion formation in MALDI mass spectrometry. Mass Spectrom Rev 17(5):337–366
6. Khatib-Shahidi S et al (2006) Direct molecular analysis of whole-body animal tissue sections by imaging MALDI mass spectrometry. Anal Chem 78(18):6448–6456
7. Passarelli MK, Ewing AG (2013) Single-cell imaging mass spectrometry. Curr Opin Chem Biol 17(5):854–859
8. Amantonico A et al (2008) Mass spectrometric method for analyzing metabolites in yeast with single cell sensitivity. Angew Chem Int Ed 47 (29):5382–5385
9. Krismer J et al (2015) Population screening of Chlamydomonas reinhardtii with single-cell resolution using a high-throughput micro scale sample preparation for MALDI mass-spectrometry. Appl Environ Microbiol 81
10. Harris EH, Stern D, Witman G (2009) Chapter 8—Chlamydomonas in the laboratory. In: Harris EH (ed) The Chlamydomonas sourcebook, 2nd edn. Academic Press, London, pp 241–302
11. Rodolfi L et al (2009) Microalgae for oil: strain selection, induction of lipid synthesis and outdoor mass cultivation in a low-cost photobioreactor. Biotechnol Bioeng 102(1):100–112
12. Harris EH, Stern D, Witman G (2009) Chapter 2—Cell architecture. In: Harris EH, Stern D, Witman G (eds) The Chlamydomonas sourcebook, 2nd edn. Academic Press, London, pp 25–64
13. Strupat K, Karas M, Hillenkamp F (1991) 2,5-Dihydroxybenzoic acid: a new matrix for laser desorption—ionization mass spectrometry. Int J Mass Spectrom Ion Process 111:89–102

Chapter 10

Sample Preparation and Analysis of Single Cells Using High Performance MALDI FTICR Mass Spectrometry

Bo Yang, Tina Tsui, Richard M. Caprioli, and Jeremy L. Norris

Abstract

Imaging mass spectrometry is a powerful technology that combines the molecular measurements of mass spectrometry with the spatial information inherent to microscopy. This unique combination of capabilities is ideally suited for the analysis of metabolites and lipids from single cells. This chapter describes a methodology for the sample preparation and analysis of single cells using high performance MALDI FTICR MS. Using this approach, we are able to generate profiles of lipid and metabolite expression from single cells that characterize cellular heterogeneity. This approach also enables the detection of variations in the expression profiles of lipids and metabolites induced by chemical stimulation of the cells. These results demonstrate that MALDI IMS provides an insightful view of lipid and metabolite expression useful in the characterization of a number of biological systems at the single cell level.

Key words MALDI, Mass spectrometry, FTICR, Single cell analysis, Sample preparation, Metabolomics, Lipidomics

1 Introduction

As we seek to more thoroughly understand the complexity of the living cell and the molecular events responsible for the maintenance of health and disease, the molecular makeup of individual cells and their interconnected role in the context of tissue and organs represents a gap in our current knowledge of biology [1]. Individual cells that exist as part of a population of cells frequently express different phenotypes, caused by phenomena such as cell cycle alterations and cellular differentiation, even in populations of cells that appear homogeneous [2]. In order to gain a full understanding of this heterogeneity of cell phenotypes, one must understand the biology of the cell in its native environment. The comprehensive molecular analysis of single cells is a formidable challenge where many thousands of biomolecules orchestrate the living functions of that cell. Among the available commercial analytical technologies, no single platform is suited to performing such analyses in an unbiased

Bindesh Shrestha (ed.), *Single Cell Metabolism: Methods and Protocols*, Methods in Molecular Biology, vol. 2064, https://doi.org/10.1007/978-1-4939-9831-9_10,

fashion since they lack the spatial resolution and the sensitivity required to measure the amounts of most biomolecules in the single cell. Sequencing technologies now provide the throughput and sensitivity required to measure gene and transcription-level variation at the single cell level [3, 4]. Though less developed, recent advances in proteomics technologies have opened up single cell analyses of proteins [5]. However, large-scale untargeted evaluations of other important biomolecules that delineate the functional phenotype of the cell (*e.g.*, lipids and metabolites) remain elusive and are not addressed by such approaches. Likewise, targeted approaches such as immunohistochemistry (IHC) address a relatively small number of proteins, for which specific antibodies are available, but often cannot be adapted to measure most lipids and metabolites. The absence of sensitive analytical tools and methods for single cell lipidomics and metabolomics will result in a significant deficit in our understanding of the role of these analytes in human disease.

Matrix-assisted laser desorption/ionization imaging mass spectrometry (MALDI IMS) is a technology that uniquely combines the sensitivity and molecular specificity of mass spectrometry with the important spatial information provided by microscopy [6]. This technology utilizes a laser to desorb and ionize molecules from specific cells on a microscope slide so that they can be analyzed using a mass spectrometer. Cells can be targeted for IMS analysis guided by microscopy images, or the entire specimen is analyzed by regular sampling of the whole surface, collecting mass spectra at each raster location. The molecular information provided by mass spectrometry, combined with spatial information, can be used to reconstruct images of the sample based on unique molecules or can define specific molecular phenotypes based on the measured molecular expression. This technology has the advantage of being label-free, allowing unbiased analysis of metabolites and lipids directly from cells and tissues. Additionally, this MS platform provides the throughput to analyze many individual cells to elucidate differences in molecular expression in cells that are nominally of the same type [7–9]. However, this technology has not yet been widely applied to the analysis of single cells from human cells and tissues because of limited sensitivity and spatial resolution.

Recent technological advancements in MALDI instrumentation have significantly increased the sensitivity and throughput while still retaining the molecular specificity of the IMS approach. Modern instrumentation frequently operates at acquisition rates exceeding 1 kHz, making it possible to collect many thousands of spectra per hour. Laser optical systems can be modified to enable IMS at dimensions below that of a single cell [10–12]. High field FTICR instruments enable unprecedented analyte sensitivity and dynamic range, providing the performance needed to monitor 100s to 1000s of low molecular weight analytes in a single analysis. This combination of performance enhancements now makes MALDI FTICR an ideal platform for single cell analysis. This study describes the preparation and analysis of single cells for analysis using MALDI FTICR MS to elucidate the unique molecular phenotype of these cells as part of a multicellular population.

2 Materials

2.1 Cell Handling and Sample Preparation

1. Microscope glass slides (3 in. × 1 in.) (*see* **Note 1**).
2. Lipopolysaccharide (LPS).
3. Phosphate-buffered saline (PBS) solution.
4. 50 mM ammonium formate solution stored at 4 °C.
5. Matrix: 1,5-Diaminonaphthalene (DAN).
6. Filter paper (7 cm).
7. Matrix recrystallization solution: MilliQ water: Chloroform: Methanol = 4:3:1 (v/v/v).
8. Mouse RAW 264.7 Cells: This cell line was established from a tumor induced by Abelson murine leukemia virus.

2.2 Mass Spectrometry Analysis

1. Calibration standard: electrospray ionization tuning mix (Agilent Technologies, Santa Clara, CA, USA).

2.3 Post-Analysis H&E Staining

1. Ethanol solution A: 70% ethanol/MilliQ water (v/v).
2. Ethanol solution B: 95% ethanol/MilliQ water (v/v).
3. Ethanol solution C: 100% ethanol/MilliQ water (v/v).
4. Hematoxylin solution:

   1 g Hematoxylin,

   45.93 g Aluminum potassium sulphate (dodecahydrate),

   379 mL Distilled/deionized water,

   100 mL Glycerol,

   0.1 g Sodium iodate.
5. Eosin solution:

   50 mL Eosin Y solution, Intensified, histology grade,

   1 g Phloxine B,

   100 mL Distilled/deionized water.
6. Xylene: Histological grade.
7. Slide Preparation Mounting Medium: Xylene-based.
8. Microscope cover slips (1 in. × 1 in.).

2.4 Instrumentation

1. Leica slide scanner SCN400 (Leica Microsystems Inc., Buffalo Grove, IL, USA).
2. Sublimation apparatus for application of matrix [13].
3. MALDI mass spectrometer: Bruker Daltonics solariX 15 Tesla MALDI-Fourier transform ion cyclotron resonance (FT-ICR) mass spectrometer.

2.5 Software

1. MALDI Mass Spectrometer control software for setting up experiments and acquisition of mass spectra: *ftmsControl 2.0* (Bruker Daltonics Inc., Billerica, MA, USA).
2. MALDI Image acquisition and visualization: *FlexImaging 4.0* (Bruker Daltonics Inc., Billerica, MA, USA).
3. *DataAnalysis 4.2* (Bruker Daltonics Inc., Billerica, MA, USA) allows for extraction and conversion of mass spectra to XML or ASCII file formats. These files can then be imported into other commercial software such as Excel, MATLAB, and R for further data processing.
4. *R 3.2.0* (The R Project for Statistical Computing, http://www.r-project.org) enables data processing and analysis such as statistical analysis and data clustering and classification.
5. *OriginPro 7* (OriginLab Inc., Northampton, MA, USA) was used to generate plots for presentation and publication of the data.

3 Methods

3.1 Cell Handling and Sample Preparation

1. Normal (no stimulation) RAW 264.7 cells are plated on microscopy glass slides at 1.5×10^6 cells/flask for 4 h (*see* **Note 2**).
2. LPS stimulation: After 4 h, one flask of cells is stimulated by exposure to 100 ng/mL LPS in PBS buffer for 16 h in culture, whereas the other cells keep growing at the normal conditions (*see* **Note 3**).
3. After plating, the cells are adherent to the slides and are washed with PBS buffer twice to remove cell culture media (*see* **Note 4**). The cell slides are snap-frozen in liquid nitrogen and stored at −80 °C until analysis.
4. On the day of analysis, the slides are stored under vacuum in a desiccator for 30 min and allowed to thaw. This approach reduces moisture condensation during thawing of the cells and thereby reduces possible delocalization of analytes on the cells.
5. The slides are then rinsed by sequentially dipping the cell slides into a Petri dish filled with 50 mM cold ammonium formate for

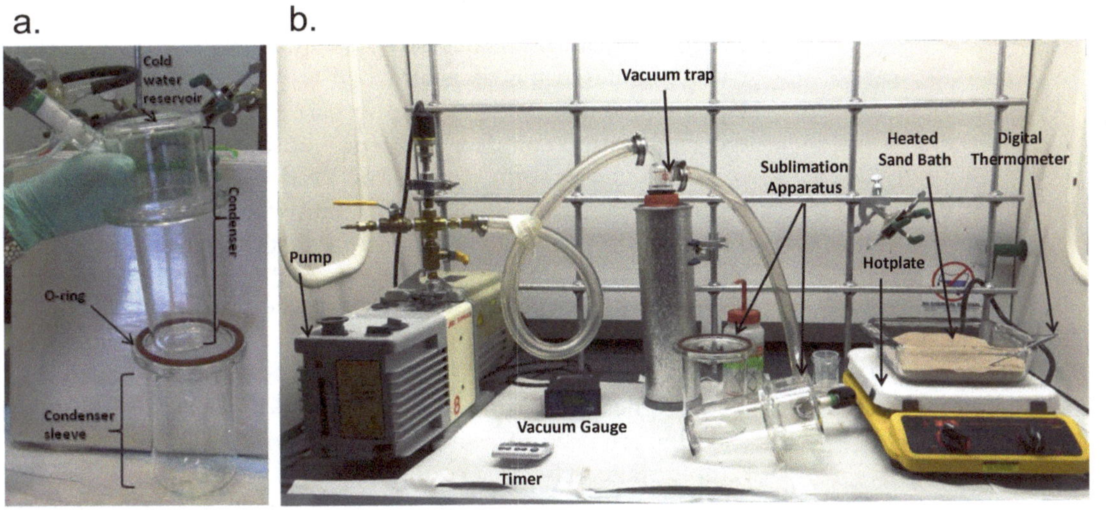

Fig. 1 (**a**) Schematic of the sublimation apparatus. The slide is fixed to the bottom of the condenser using metallic tape and the MALDI matrix is placed in the bottom of the condenser sleeve. The condenser is placed into the sleeve such that the sample is inverted over the matrix powder. (**b**) Schematic diagram of the sublimation equipment setup. The assembled condenser (**a**) is placed into the heated sand bath and the vacuum is applied to initiate matrix sublimation

30 s to reduce salts from the cells and increase signal intensities of lipids and metabolites [14]. Excess ammonium formate is quickly removed by wiping the slide edges with a Kimwipe.

6. After rinsing, the cell slides are stored in a desiccator for 30–60 min to allow solvent to dry.
7. Fiducial points are placed on the slides before image scanning (*see* **Note 5**). Scanning is performed before application of matrix to obtain a clear image of the cells. Slide images are obtained at 40× magnification using a Leica Image Scanner. The scanned slide image is converted to JPEG format.
8. Application of matrix is performed using a custom-built sublimation apparatus (Fig. 1). This approach has been generally described in the literature for tissue IMS [13] and adapted for single cell analysis as described. The core condenser sleeve was purchased from Chemglass Life Sciences and fits a plate diameter of 70 mm. About 100 mg MALDI matrix was placed at the bottom of the condenser sleeve. Pressure of the condenser sleeve was held at <50 mTorr by a rough pump and monitored by a digital thermocouple vacuum gauge controller. The condenser sleeve was placed on a sand bath heated by a hot plate (Corning 6795-200). The temperature was monitored by a digital thermometer. For application of DAN, the MALDI matrix used in the analysis of RAW 264.7 cells, the condenser sleeve was held in 95 °C sand bath for approximately 9 min (*see* **Note 6**).

9. Matrix rehydration: Matrix rehydration is performed to enhance signal intensities of lipids and metabolites. The slide with sublimated matrix is attached to a copper plate, which is taped to the underside of the top part of the Petri dish. Matrix recrystallization solution is pipetted onto a piece of filter paper placed in the bottom part of the Petri dish. The Petri dish is reassembled and sealed using autoclavable tape (Fisher, 15-901-30B) to create a hydration chamber. The Petri dish is left on an oven preheated to 85 °C for 2 min, and then the top is removed and the slide is allowed to dry in the oven for 1 min.

3.2 MALDI-Imaging Mass Spectrometry (IMS) of Single Cells

1. The MALDI/ESI-FTICR mass spectrometer is mass calibrated using Agilent ESI tuning mix. Fiducial registration points on the slide are used to calibrate the *X* and *Y* coordinates on the sample stage. The area of slide to be analyzed is annotated in *FlexImaging* software, and information related to the slide geometry is sent to *ftmsControl* software automatically. Laser raster size is defined in the *FlexImaging* software based on size of the cells and signal intensity. For each position, one spectrum is acquired by summing up 300–500 laser shots.
2. After calibration of the instrument, the selected area is analyzed by a 15 T MALDI-FTICR mass spectrometer coupled to a 355 nm Nd:YAG laser and operated in negative mode in the mass range of 300–2000 Da.

3.3 Matrix Removal/Staining

1. After mass spectrometry measurements are completed, the matrix is removed by dipping the slides into 100% acetone solution for 5 s.
2. Post-analysis H&E staining protocol: The slides are successively immersed in ethanol solutions B, A, and MilliQ water for 30 s each, hemotoxylin solution for 2 min, MilliQ water 20 s, Ethanol solution A and B for 30 s each, eosin solution for 1 min, ethanol solution B and C for 30 s each, xylene for 2–3 min. The analyzed region are covered with cover slips and allowed to dry.
3. The acquired data can be visualized in *FlexImaging*, whereas basic preprocessing is carried out in *DataAnalysis* software (Fig. 2).

3.4 Data Processing

1. Guided by H&E stained image, mass spectra are extracted from normal and LPS stimulated single cells using *FlexImaging* and *DataAnalysis* software.
2. Each mass spectrum is exported as ASCII file containing column delimited mass list and ion intensities.
3. XY files for all single cells are imported using scripts written in *R*.

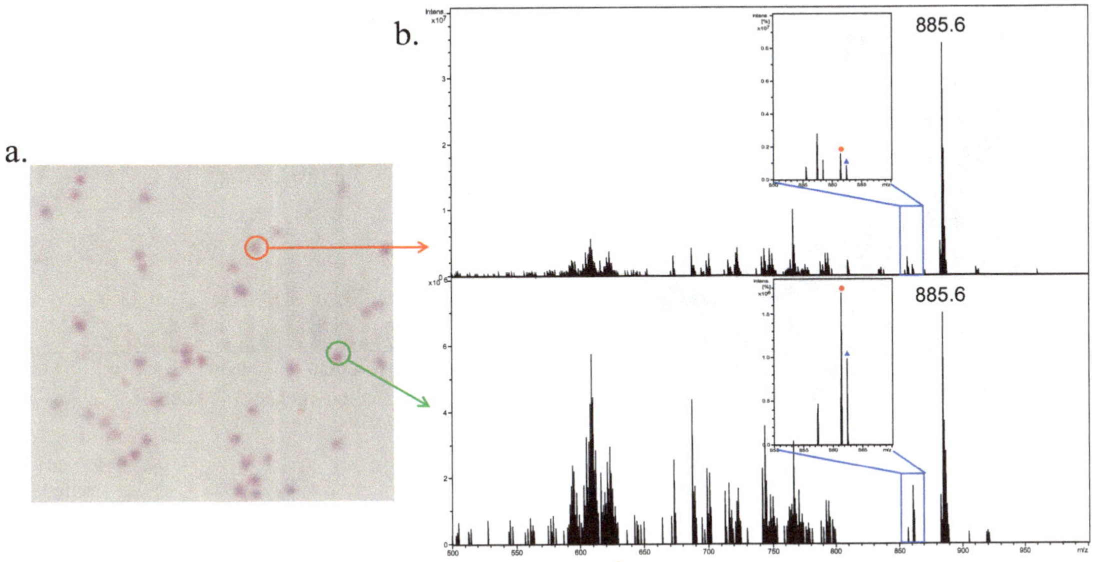

Fig. 2 (**a**) H&E stained image. (**b**). Representative mass spectra of two single cells highlighted by red and green circles in the H&E stained image, respectively. The original spectra in the blue box are enlarged to the same relative intensity range (0–25%) as compared to the base peak ($m/z = 885.6$) for better visualization. Ions corresponding to $m/z = 861.6$ (red circle●) and 862.6 (blue triangle▲) exhibit different relative intensities relative to the base peak

4. Ion intensities are normalized to total ion current (TIC).
5. Peak picking is performed using R scripts. The minimum intensity threshold is set to 1.2% of the base peak intensity. Similar peaks within a 15 ppm mass window are grouped together and an average mass is calculated to represent the group. The processed results are exported as an Excel file containing a mass list and ion intensities for each single cell spectrum.
6. Histograms of absolute ion intensities are generated and compared for the normal and LPS stimulated single cells, and representative graphs are shown in Fig. 3.

3.5 Concluding Remarks and Future Directions

Much of what is known about cellular metabolism and lipid function in the cell has been gleaned from studies of samples derived from mixtures of cells. These average measurements taken from entire cell populations have been enormously valuable and have greatly contributed to our current understanding of cellular processes. In spite of this progress, there exists a limit to our understanding of living systems that can only be overcome by developing the capabilities to study molecular expression at the level of a single cell. The advanced mass spectrometry instrumentation described in this study provides a powerful opportunity to probe single cells and to characterize their complex roles in the context of living systems. The methodology described, utilizing MALDI FTICR IMS, was

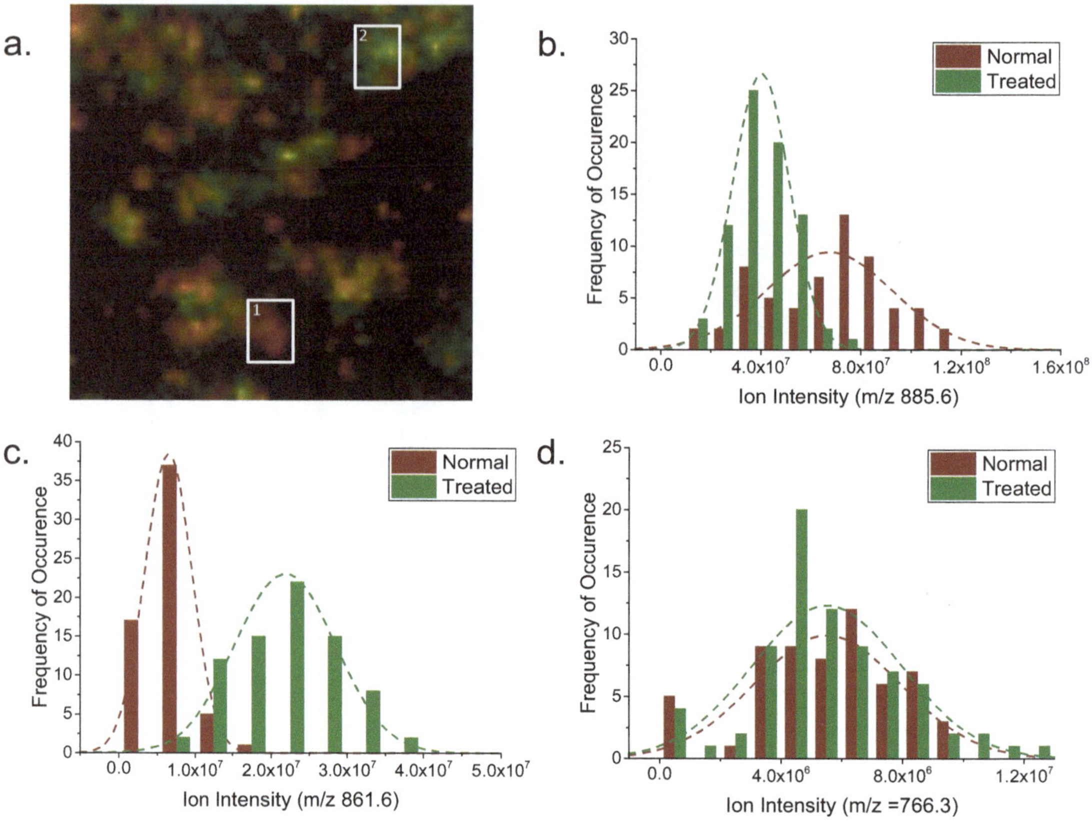

Fig. 3 (**a**). Overlaid ion images of $m/z = 766.5$ (green) and 885.6 (red) from normal RAW 264.7 cells demonstrate heterogeneous molecular expression of each molecular species. For example, region 1 shows a group of cells expressing a relative increase of ion 885.6 and region 2 shows a relative increase in ion 766.5. (**b–d**). Histograms of ion intensities of $m/z = 885.6$ and 861.6, and 766.3 from normal ($n = 60$) and LPS stimulated ($n = 70$) RAW 264.7 cells. These results show that LPS stimulation selectively influences the expression of selected molecular species (**b**) *m/z* 885.6 and (**c**) *m/z* 861.6, shifting both the position of the mean and altering the shape of the distribution of ion abundance. The ion at *m/z* 766.3 shows no evidence of altered expression upon LPS stimulation

demonstrated to have sufficient sensitivity, spatial resolution, and specificity for single cell lipidomics and metabolomics. Further applying this technology to different cellular states, we have measured variations (narrowed, widened, shifted distributions) in the expression profiles of these lipids and metabolites upon chemical stimulation of the cells.

4 Notes

1. Standard borosilicate glass microscope slides are suitable for MALDI analysis using ion trapping instruments, like those used in this study, or quadrupole time-of-flight instruments.

When adapting the sample preparation approach described here to be used for sample analysis on time-of-flight instruments, conductive slides must be used. The authors recommend the use of indium tin oxide (ITO) coated slides, which can be purchased in the same 3 in. × 1 in. microscope slide format from a number of commercial sources.

2. For single cell analysis, the cell culture conditions would ideally result in a large amount of spatially separated single cells. Therefore, the cell plating protocol is optimized based on cell types to obtain ideal cellular distribution. The major factors that influence cellular distribution are the original cell counts and plating time. Because the ammonium formate washing can remove a small amount of cells from the slides, it is better to plate cells at a slightly higher density than the ideal cell distribution. Stimulation of the cells and changes to the growth conditions will alter cell growth doubling time, and hence require re-optimization of the cell plating protocol.
3. LPS was selected as a chemical stimulant to provide a means to evaluate the analytical capabilities of MALDI IMS and the sensitivity of the method to observe alterations in molecular expression. LPS is a potent activator of monocytes and macrophages. LPS stimulates immune responses by interacting with the membrane receptor CD14 to induce the generation of cytokines such as tumor necrosis factor (TNF)-a, interleukin (IL)-1, and IL-6 [15].
4. Complete removal of the cell culture media is required for the analysis. Lipid, metabolite, and protein components of the cell culture media will interfere with the MALDI analysis of the endogenous compounds of interest. For each growth condition, this media removal step should be tested to ensure complete removal of the background contribution from the cell culture media.
5. Fiducial points need to be as fine as possible to achieve the most accurate registration of the *X*, *Y* coordinates. Fiducial points can be placed using ultrafine Sharpie markers or printed grid stickers (e.g., Electron Microscopy Sciences, 70366-14). To ensure precise overlay of IMS ion image with H&E stained image, fiducial points must survive the matrix removal and H&E staining procedures.
6. Optimization of the sublimation protocol, such as sand bath temperature and duration of heat application should be performed based on the specific MALDI matrix used and the class of analytes of interest.

Acknowledgments

This work is supported by NIH grants R33DK105149 and P41GM103391. The authors also acknowledge Nichole A. Lobdell from Eric P. Skaar Lab for providing the cultured cells for this work.

References

1. Zenobi R (2013) Single-cell metabolomics: analytical and biological perspectives. Science 342(6163):1243259. https://doi.org/10.1126/science.1243259
2. Kolitz SE, Lauffenburger DA (2012) Measurement and modeling of signaling at the single-cell level. Biochemistry 51(38):7433–7443. https://doi.org/10.1021/bi300846p
3. Kalisky T, Quake SR (2011) Single-cell genomics. Nat Methods 8(4):311–314. https://doi.org/10.1038/nmeth0411-311
4. Tang F, Lao K, Surani MA (2011) Development and applications of single-cell transcriptome analysis. Nat Methods 8(4 Suppl):S6–S11. https://doi.org/10.1038/nmeth.1557
5. Shapiro E, Biezuner T, Linnarsson S (2013) Single-cell sequencing-based technologies will revolutionize whole-organism science. Nat Rev Genet 14(9):618–630. https://doi.org/10.1038/nrg3542
6. Norris JL, Caprioli RM (2013) Analysis of tissue specimens by matrix-assisted laser desorption/ionization imaging mass spectrometry in biological and clinical research. Chem Rev 113(4):2309–2342. https://doi.org/10.1021/cr3004295
7. Li L, Garder RW, Sweedler JV (2000) Single-cell MALDI: a new tool for direct peptide profiling. Trends Biotechnol 18(4):151–160. https://doi.org/10.1016/S0167-7799(00)01427-X
8. Amantonico A, Urban PL, Fagerer SR, Balabin RM, Zenobi R (2010) Single-cell MALDI-MS as an analytical tool for studying intrapopulation metabolic heterogeneity of unicellular organisms. Anal Chem 82(17):7394–7400. https://doi.org/10.1021/ac1015326
9. Shrestha B, Vertes A (2009) In situ metabolic profiling of single cells by laser ablation electrospray ionization mass spectrometry. Anal Chem 81(20):8265–8271. https://doi.org/10.1021/ac901525g
10. Zavalin A, Yang J, Haase A, Holle A, Caprioli R (2014) Implementation of a Gaussian beam laser and aspheric optics for high spatial resolution MALDI imaging MS. J Am Soc Mass Spectrom 25(6):1079–1082. https://doi.org/10.1007/s13361-014-0872-5
11. Zavalin A, Yang J, Caprioli RM (2013) Laser beam filtration for high spatial resolution MALDI imaging mass spectrometry. J Am Soc Mass Spectrom 24(7):1153–1156. https://doi.org/10.1007/s13361-013-0638-5
12. Zavalin A, Todd EM, Rawhouser PD, Yang J, Norris JL, Caprioli RM (2012) Direct imaging of single cells and tissue at sub-cellular spatial resolution using transmission geometry MALDI MS. J Mass Spectrom 47(11):1473–1481. https://doi.org/10.1002/jms.3108
13. Hankin JA, Barkley RM, Murphy RC (2007) Sublimation as a method of matrix application for mass spectrometric imaging. J Am Soc Mass Spectrom 18(9):1646–1652. https://doi.org/10.1016/j.jasms.2007.06.010
14. Angel PM, Spraggins JM, Baldwin HS, Caprioli RM (2012) Enhanced sensitivity for high spatial resolution lipid analysis by negative ion mode matrix assisted laser desorption ionization imaging mass spectrometry. Anal Chem 84(3):1557–1564. https://doi.org/10.1021/ac202383m
15. Meng F, Lowell CA (1997) Lipopolysaccharide (LPS)-induced macrophage activation and signal transduction in the absence of Src-family kinases Hck, Fgr, and Lyn. J Exp Med 185(9):1661–1670

Chapter 11

Toward Single Cell Molecular Imaging by Matrix-Free Nanophotonic Laser Desorption Ionization Mass Spectrometry

Sylwia A. Stopka and Akos Vertes

Abstract

In recent years, innovations in mass spectrometry imaging (MSI) have enabled simultaneous detection and mapping of biomolecules and xenobiotics directly from biological tissues and single cells. Matrix-assisted laser desorption ionization (MALDI) has been the most widely embraced MSI technique. However, this technique can exhibit ion suppression effects hindering metabolite coverage and possesses a narrow dynamic range. Nanophotonic platforms, e.g., silicon nanopost array (NAPA) structures, can be used as an alternative for matrix-free imaging of biological tissues. Here, we present a protocol for MSI of large and small adherent cell clusters by laser desorption ionization from NAPA with minimal sample preparation.

Key words Mass spectrometry imaging, Cell culture, Cell clusters, Metabolites, Nanopost array, NAPA, Molecular imaging

1 Introduction

The stochastic nature of gene and protein expression results in cellular heterogeneity within complex biological tissues, and can also manifest in fluctuations of metabolite levels [1]. To capture a true snapshot of the molecular composition of a single cell with minimum perturbation, technological advancements are needed. Mass spectrometry (MS) is a promising tool for the detection, quantification, and identification of an array of biomolecules. However, general MS measurements of biological tissues include bulk measurements that mask individual cell signals, or require sample preparation, which obscures the spatial information for molecular distributions. To overcome these challenges, MS imaging (MSI) techniques have been introduced based on, for example, matrix-assisted laser desorption ionization (MALDI) [2, 3]. Over the years, improvements in sensitivity and lateral resolution have made it possible to perform tissue imaging with cellular resolution

Bindesh Shrestha (ed.), *Single Cell Metabolism: Methods and Protocols*, Methods in Molecular Biology, vol. 2064, https://doi.org/10.1007/978-1-4939-9831-9_11, © Springer Science+Business Media, LLC, part of Springer Nature 2020

and capture the spatial distributions for an array of biomolecules, e.g., metabolites, lipids, proteins, and xenobiotics, from single cells [4, 5]. However, spectral interferences and ion suppression are sometimes observed due to the application of the matrix, hindering small molecule analysis [6]. Quantitation by MALDI-MSI also remains a challenge because of the limited dynamic range of the MALDI signal.

Several emerging matrix-free platforms have demonstrated successful MSI of small molecules from biological tissues using laser deposition ionization LDI [7]. These ionization sources rely on the interaction between the laser radiation and a nanostructure to induce LDI. Among these only a few have been demonstrated for MSI, including desorption ionization on porous silicon (DIOS) [8], nanostructure initiator mass spectrometry (NIMS) [9–11] and silicon nanopost array (NAPA) platforms [12–14]. Due to the exquisite sensitivity and dynamic range exhibited by these ionization sources for small molecule analysis, detection of metabolites in single yeast cells was achieved [15, 16]. Furthermore, these silicon structures can be a promising substrate for culturing cells and microfluidics applications.

Over the years, biocompatible materials have gained increasing interest. For example, porous silicon possesses unique chemical and physical properties that are conducive for use as a substrate for cell growth [17, 18]. Silicon is often used in biosensors, as a substrate in lab-on-a-chip designs or as tissue engineering scaffolds [19–22]. Silicon nanostructure arrays have also been used to investigate cellular morphology, and can be tuned to capture circuiting tumor cells [23, 24].

Adherent mammalian cells can be a model system to study human diseases, drug interactions, and cellular heterogeneity in situ [25–27]. However, current practices for handling these cells include cell lysis and enzymatic digestion, which may alter their biochemical composition and rely on bulk measurements. In this contribution, we present a protocol that uses silicon NAPA as a platform for culturing the cells and MSI with minimal cell manipulation.

2 Materials

2.1 Reagents and Chemicals

1. Cell culture medium consists of Eagle's Minimum Essential Medium (EMEM) (30-2003; ATCC, Manassas, VA) supplemented with 10% fetal bovine serum (FBS) (30-2020; ATCC, Manassas, VA) and 1% Penicillin-streptomycin (ThermoFisher, Waltham, MA).
2. Trypsin-EDTA (0.25%) (ThermoFisher, Waltham, MA) for cell detachment.

3. Trypan blue solution (0.4%) (ThermoFisher, Waltham, MA) for cell counting.
4. 70% ethanol can be used to sterilize the NAPA chip.

2.2 Biological Samples and Equipment

1. Cryogenic stock of human hepatocarcinoma cells (HepG2/C3A; CRL-10741, ATCC, Manassas, VA) derived from liver (*see* **Note 1**).
2. A humidity-controlled cell incubator (HeraCell; Thermo Scientific, Waltham, MA) that operates at a constant 5% CO_2 and 37 °C.
3. Automated cell counter (Countess; Invitrogen, Eugene, OR).
4. 35 cm diameter polystyrene Petri culture dishes.
5. Optical microscope with DIC option and scanning electron microscope.

2.3 NAPA Imaging Chips

1. Patterning for nanofabrication of the NAPA substrate is performed on low resistivity p-type silicon wafers by either e-beam lithography (EBL) or deep UV projection lithography (DUV-PL). The former is a much lengthier process and is not conducive to making large area NAPA imaging chips.
2. Both fabrication methods of the silicon NAPA are described in detail in our previous work [12, 28].
3. Briefly, for DUV-PL a layer of SiO_2 is developed by thermal oxidation followed by spin coating a negative tone photoresist. The wafer then undergoes DUV-PL exposure at 248 nm wavelength to create the nanopost pattern.
4. The photoresist is then developed and the exposed SiO_2 is etched away to form a hard mask for the nanoposts.
5. The remaining photoresist is striped, leaving only a layer of SiO_2 hard mask protecting the positions of the nanoposts. Using a mixture of etchant gases, the silicon surrounding the nanoposts is etched away in a deep reactive ion etching (DRIE) process.
6. The fabricated nanoposts are 1000 nm in height, 150 nm in diameter, and 337 nm in periodicity. The NAPA imaging chips were produced to a size of 4×4 mm^2 to accommodate biological tissue sections.

2.4 Mass Spectrometer and Supplies

1. Oil-free vacuum desiccator system (813-600; Ted Pella, Inc., Redding, CA) to dry the sample on the NAPA chip prior to analysis.
2. Dry box for NAPA storage.
3. Double-sided carbon tape to mount the chips onto a MALDI plate (*see* **Note 2**).

4. A linear ion trap-orbitrap hybrid mass spectrometer fitted with a MALDI ion source (MALDI LTQ-Orbitrap XL; Thermo Fisher Scientific Inc., Bremen, Germany) can be used for NAPA-LDI-MSI experiments. Ion assignments are facilitated by tandem MS (MS/MS). The nitrogen laser within the source operates at a wavelength and repetition rate of 337 nm and 60 Hz, respectively.

2.5 Software

1. Xcalibur (3.0.63) (ThermoFisher Scientific Inc., Bremen, Germany) for the interpretation of mass spectra.
2. ImageQuest (1.1.0) (ThermoFisher Scientific Inc., Bremen, Germany) for molecular image processing and the reconstruction of chemical images.

3 Methods

3.1 Culturing of Mammalian Cells on NAPA Substrates

1. A stock of HepG2/C3A cells should be maintained by culturing them in a T75 flask in the supplemented EMEM medium. Grow the cells till 70–80% confluence. To detach the cells from the flask surface, aspirate the medium out completely, wash with 1× PBS, aspirate the PBS solution, and add 4 mL of pre-warmed trypsin EDTA (Fig. 1(1)).
2. Place the flask into the humidity-controlled incubator for 10 min for the digestion of the extracellular matrix to be complete; tap the flask gently to insure cell detachment.
3. Deactivate the trypsin with the addition of fresh pre-warmed EMEM media and centrifuge (1000 × *g*) the cell suspension to form a cell pellet (Fig. 1(2)). Aspirate the trypsin and medium solution from the cells and resuspend the cell pellet with 2 mL of fresh medium.
4. To avoid cell clumping, gently aspirate the cell suspension up and down ten times with a pipette.
5. Measure the cell density by mixing, the cell suspension and trypan blue solution 1:1 (v/v) and use a cell counter to measure the cell numbers. Dilute the cell suspension to ~5×10^4 cells/mL.
6. Sterilize a 4×4 mm^2 NAPA chip with 70% ethanol and place the chip in a 35 mm culture dish.
7. Pipet 15 μL of the diluted cell suspension onto the NAPA chip, resulting in a seeding density of ~750 cells per 4×4 mm^2 NAPA chip (Fig. 1(3)).
8. Place the NAPA chip covered by the cells in the incubator for 15 min to allow the cells to adhere to the silicon substrate.

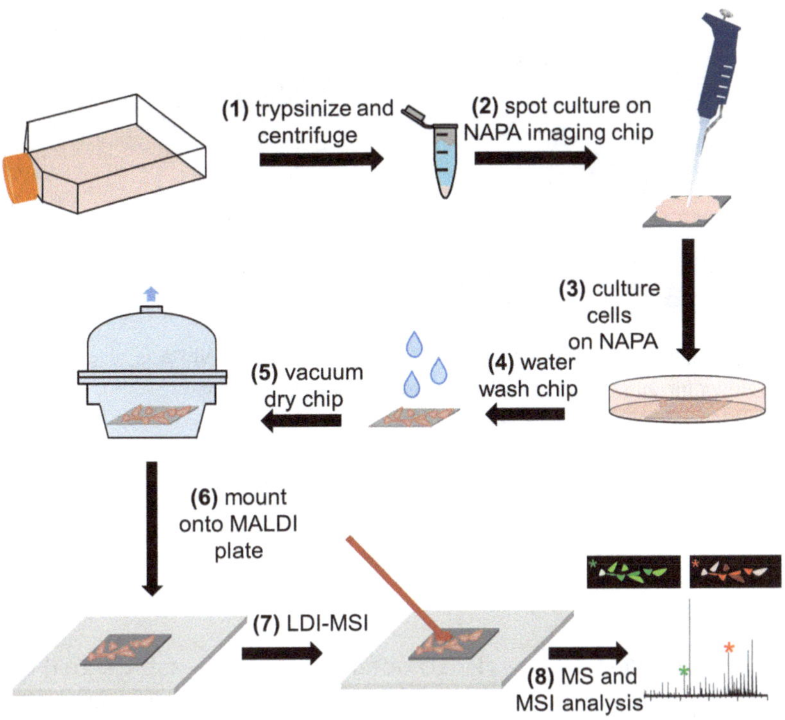

Fig. 1 Workflow of sample preparation and MSI analysis for mammalian cells cultured on NAPA imaging chips. A flask of hepatocarcinoma cells is first (1) trypsinized to detach the cells from the flask followed by centrifugation to form a cell pellet. (2) The cell pellet is then dispensed onto the NAPA structure and (3) cultured for six days. Once the cells are ready for analysis, the NAPA chip is (4) water washed, (5) vacuumed dried, and (6) affixed onto a standard MALDI plate for MSI. (7) Imaging experiments are performed by rastering over the cultured cells with a laser with mass spectra acquired at every pixel. Molecular images (8) are constructed from the MSI experiment by plotting ion intensity distributions for biomolecules directly from the cultured cells

9. Gently fill the 35 mm dish containing the NAPA chips with ¾ EMEM medium and place them in the incubator. Change out the medium every other day and monitor the cell growth daily using optical microscopy (Fig. 1(3)).

3.2 Sample Preparation for LDI-MSI

1. After 6 days of culturing the cells on the NAPA structure, remove the chips from the Petri dish with tweezers. Rotate the chip to vertical orientation and gently wash it with water to remove any excess media (*see* **Note 3**) (Fig. 1(4)).
2. Place the NAPA chips containing the cells inside an oil-free vacuum desiccator (at ~75 Torr pressure) for 10 min for the cells to dry on the NAPA substrate (Fig. 1(5)).
3. Throughout culturing, washing, and vacuum drying, the cells should be inspected using optical microscopy to insure that

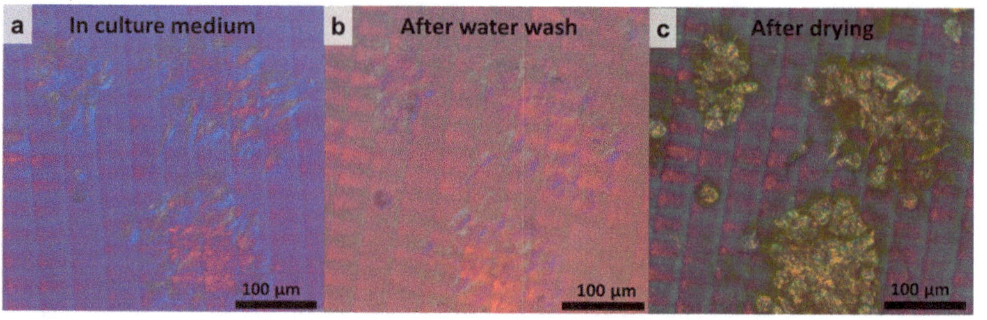

Fig. 2 Optical images of the HepG2C3A cells cultured on NAPA substrate in (**a**) culturing medium, (**b**) after a gentle water wash, and (**c**) after drying the cells for analysis

their morphology is unchanged and no physical damage has been inflicted (Fig. 2a–c).

4. Once the NAPA chip is dry, using a double-sided carbon tape, affix it to a standard MALDI plate (Fig. 1(6)) and insert it into the mass spectrometer for LDI-MSI acquisition (Fig.1(7, 8)).

3.3 Optimization of NAPA-MSI Signal

1. To obtain robust and stable signal during the MSI run, several parameters need to be optimized on a test NAPA chip that has undergone similar sample preparation steps and contains adherent cells.
2. Since these are imaging experiments and the ions are collected as discrete packets at each pixel, the automatic gain control (AGC) feature is turned off.
3. The laser energy for MSI experiments should be optimized by firing varied fluences directly on the cell cluster surface ranging from 60 to 100 mJ/cm^2 (*see* **Note 4**). Optimization of the NAPA-MSI signal is complete when its intensity is the highest and in-source fragmentation is at the minimum.
4. The number of laser shots per pixel should be configured, as well. Once the laser fluence is optimized, adjust the number of shots between 2 and 10 per pixel. Once again select the configuration that provides the best signal.
5. Lastly, for cell cluster imaging a spatial resolution of <50 µm is recommended; based on the laser spot size in the mass spectrometer, oversampling maybe required. In this protocol, the laser spot size is $100 \times 80\ \mu m^2$, thus oversampling may be needed.
6. Mass resolving power can be varied between $R = 30{,}000$ and 100,000 and its selection is dictated by the application. Longer acquisition times and loss in sensitivity can be expected at higher resolving powers.

3.4 NAPA Imaging

1. In order to obtain molecular images by LDI from NAPA, mass spectra are collected for all the pixels that are later rendered as false color images based on the ion intensities. A motorized stage translates the sample in two dimensions to obtain a mass spectrum at each pixel. Below are the steps for obtaining the molecular images.
2. Scan the region of interest, in this case the whole $4 \times 4\ mm^2$ NAPA chip, to produce the optical image needed for co-registration.
3. Input the optimal experimental parameters for the number of laser shots per pixel, laser fluence, and step size (e.g., 50 μm).
4. Start the MSI data acquisition.
5. Once the analysis is complete, place the instrument in standby mode.

3.5 Data Processing and Inspection of Cells After LDI

1. Using the Xcalibur software, all of the pixels from the NAPA area are averaged to show the overall signal quality (Fig. 3). In both positive and negative ion modes, small metabolites and lipid species are detected from the adherent HepG2/C3A cells.
2. (Optional) Using SEM, the NAPA chips are inspected before and after LDI-MSI (Fig. 4). Prior to laser irradiation, the cells, including their lamellipodia, are attached to the silicon substrate (Fig. 4a).

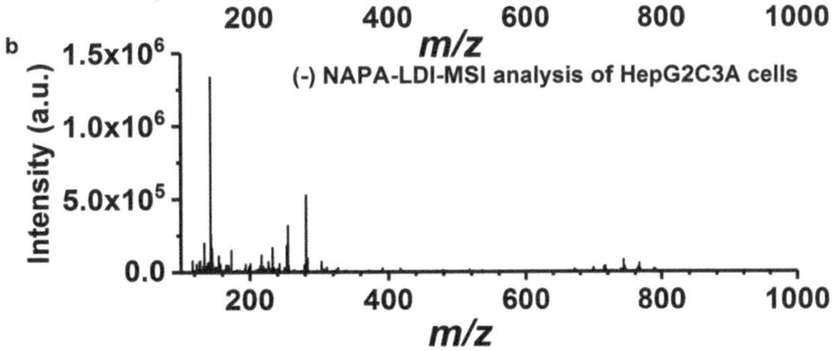

Fig. 3 Averaged mass spectra for HepG2C3A cells cultured directly on a NAPA substrate in (**a**) positive and (**b**) negative ion modes. Small metabolites and lipids are detected

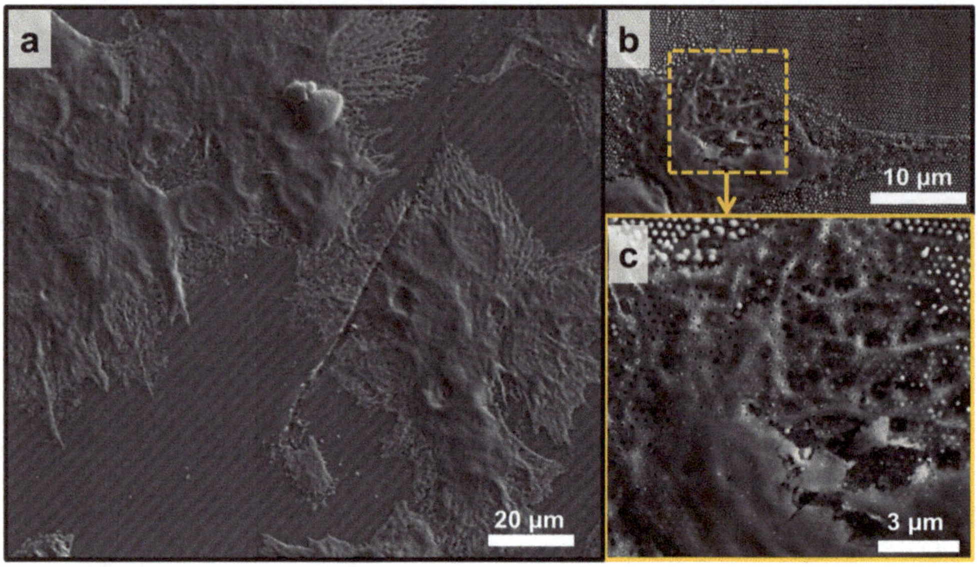

Fig. 4 (**a**) An SEM image of cells attached to the NAPA surface before LDI analysis. (**b**) After LDI analysis, SEM images revealed desorption of cellular material due to laser irradiation. (**c**) Zoomed-in region shows nanopores in the cell material that are similar in diameter and periodicity to the underlying NAPA posts

3. (Optional) After laser exposure, a zoomed in SEM image shows that cellular material is removed in the LDI process (Fig. 4b). Closer inspection of the exposed cell indicates a pattern of nanopores that follows the nanopost positions, indicating the role of nanopost heating in material desorption (Fig. 4c).

3.6 Large Cell Population Imaging

1. A cell population containing ~2000 adherent cells on a NAPA chip can be analyzed.
2. Prior to LDI-MSI, the dried NAPA chip is imaged using optical microscopy to locate the cell clusters (Fig. 5a).
3. To produce molecular images based on ion intensities, the whole NAPA chip is rastered using 50 μm step size. This results in a total of 6724 pixels to form images based on the ion intensities for particular *m/z* values at each pixel.
4. A few hundred biomolecules can be detected and their ion intensities can be spatially mapped to show their localization within the whole cell population (Fig. 5b, c). For example, the distributions of $[\text{glutathione} + K]^+$ and $[\text{cholesterol} + H\text{-}H_2O]^+$ are observed throughout the whole cell population.
5. Background ions from the NAPA chip can also be spatially mapped (*see* the blue composite image in Fig. 5d).
6. The images of these two biomolecules and the NAPA background ions can be overlaid to observe their distributions in relation to each other (Fig. 5e).

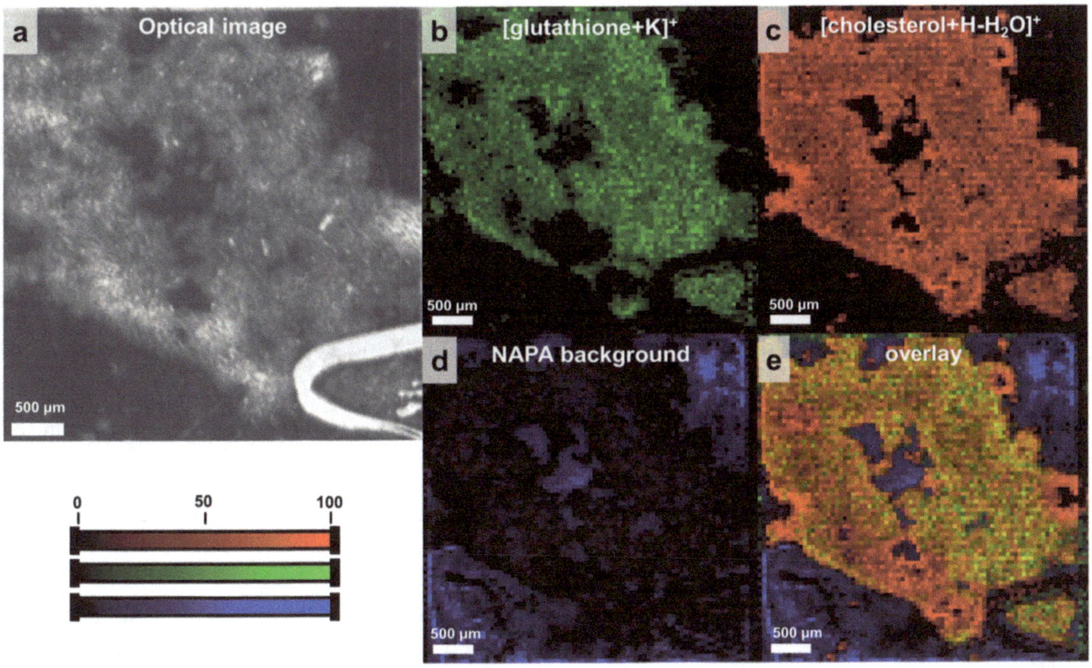

Fig. 5 (**a**) Optical image from mass spectrometer camera of NAPA imaging chip containing ~2000 HepG2/C3A cells. Molecular images based on metabolite ion intensities, e.g., (**b**) glutathione and (**c**) cholesterol can be spatially correlated with the location of cell clusters. Selection of (**d**) NAPA-related background peaks can reveal the exposed NAPA substrate. (**e**) Superimposed molecular images reveal overlaying spatial distributions from the cell clusters as well as the NAPA substrate

3.7 Small Cell Cluster Imaging

1. Regions of interest from the co-registered optical and molecular images can be selected to show the distribution of biomolecules from small cell clusters (Fig. 6).
2. Using the optical image the number of adherent cells can be counted for each small cell cluster.
3. Molecular images can be used to assess the signal and the resolution of spatial mapping based on the pixels corresponding to cell clusters.
4. Several metabolites and lipids can be localized to clusters of ~10 cells. For example, $[\text{phosphocholine}]^+$, $[\text{PE}(36{:}2) + \text{K}]^+$, $[\text{TG}(54{:}3) + \text{K}]^+$, and $[\text{glutathione} + \text{K}]^+$ were all detected in the small cell clusters (Fig. 6b–e).
5. Clusters of $n < 10$ cells result in no signal, indicating that the sensitivity of this LDI platform is reached. However, surface silanization or use of bowtie configured NAPA posts can improve ionization efficiency and could ultimately be used for single cell analysis.

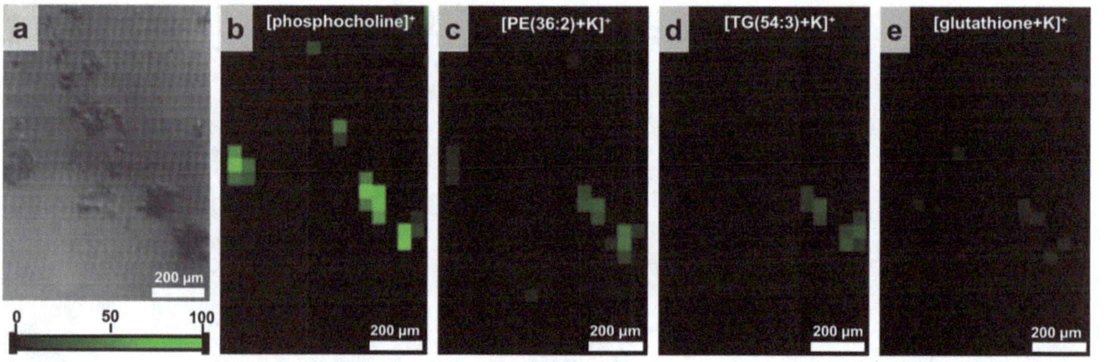

Fig. 6 (**a**) Optical image of small cell clusters attached to NAPA substrate. (**b–e**) Corresponding molecular images of small metabolites and lipid species detected from the cell clusters. Signal was only observed in cell clusters of $n = 10$ or larger

4 Notes

1. The HepG2/C3A cells require biosafety level one (BSL-1) environment and all cell handling and culturing should be performed in a BLS-1 rated hood. All equipment and instruments that may come in contact with the cells should have biohazard sign posted.
2. Depending on the type of mass spectrometer, make sure there is enough clearance for the carbon tape and NAPA chip to be inserted into the instrument.
3. When pipetting directly onto the NAPA chips, avoid touching the nanostructure with the pipette tip. Touching the surface will damage the posts, resulting in poor signal within that area. For example, in Fig. 5 the lower right corner of the chip was damaged with a pipette tip and poor signal in that area can be seen in the molecular images.
4. Due to the rapid heating of the nanoposts, in-source fragmentation of labile compounds is observed. To minimize this effect, use lower laser fluences. Direct analysis of a standard compound mixture can be used to assess the amount of in-source fragmentation for those specific compounds.

Acknowledgments

Research was sponsored by the U.S. Army Research Office and the Defense Advanced Research Projects Agency and was accomplished under Cooperative Agreement Number W911NF-14-2-0020. The views and conclusions contained in this document are those of the authors and should not be interpreted as representing the official policies, either expressed or implied, of the Army Research Office,

DARPA, or the U.S. Government. The U.S. Government is authorized to reproduce and distribute reprints for Government purposes notwithstanding any copyright notation hereon.

References

1. Taniguchi Y (2011) Quantifying *E. coli* proteome and transcriptome with single-molecule sensitivity in single cells (vol. 329, p. 533, 2010). Science 334(6055):453–453
2. Caprioli RM (2014) Imaging mass spectrometry: molecular microscopy for enabling a new age of discovery. Proteomics 14 (7–8):807–809. https://doi.org/10.1002/pmic.201300571
3. McDonnell LA, Heeren RMA (2007) Imaging mass spectrometry. Mass Spectrom Rev 26 (4):606–643. https://doi.org/10.1002/mas.20124
4. Zavalin A, Yang JH, Caprioli R (2013) Laser beam filtration for high spatial resolution MALDI imaging mass spectrometry. J Am Soc Mass Spectrom 24(7):1153–1156. https://doi.org/10.1007/s13361-013-0638-5
5. Schuerenberg M, Luebbert C, Deininger SO, Ketterlinus R, Suckau D (2007) MALDI tissue imaging: mass spectrometric localization of biomarkers in tissue slices. Nat Methods 4(5): iii–iiv
6. Ferguson CN, Fowler JWM, Waxer JF, Gatti RA, Loo JA (2014) Mass spectrometry-based tissue imaging of small molecules. In: Woods AG, Darie CC (eds) Advancements of mass spectrometry in biomedical research, vol 806. Advances in experimental medicine and biology. Springer, New York, NY, pp 283–299. https://doi.org/10.1007/978-3-319-06068-2_12
7. Bergman N, Shevchenko D, Bergquist J (2014) Approaches for the analysis of low molecular weight compounds with laser desorption/ionization techniques and mass spectrometry. Anal Bioanal Chem 406 (1):49–61. https://doi.org/10.1007/s00216-013-7471-3
8. Wei J, Buriak JM, Siuzdak G (1999) Desorption-ionization mass spectrometry on porous silicon. Nature 399(6733):243–246. https://doi.org/10.1038/20400
9. Northen TR, Yanes O, Northen MT, Marrinucci D, Uritboonthai W, Apon J, Golledge SL, Nordstrom A, Siuzdak G (2007) Clathrate nanostructures for mass spectrometry. Nature 449(7165):1033–U1033. https://doi.org/10.1038/nature06195
10. Patti GJ, Shriver LP, Wassif CA, Woo HK, Uritboonthai W, Apon J, Manchester M, Porter FD, Siuzdak G (2010) Nanostructure-initiator mass spectrometry (NIMS) imaging of brain cholesterol metabolites in Smith-Lemli-Opitz syndrome. Neuroscience 170 (3):858–864. https://doi.org/10.1016/j.neuroscience.2010.07.038
11. Greving MP, Patti GJ, Siuzdak G (2011) Nanostructure-initiator mass spectrometry metabolite analysis and imaging. Anal Chem 83(1):2–7. https://doi.org/10.1021/ac101565f
12. Walker BN, Stolee JA, Pickel DL, Retterer ST, Vertes A (2010) Tailored silicon nanopost arrays for resonant nanophotonic ion production. J Phys Chem C 114(11):4835–4840. https://doi.org/10.1021/jp9110103
13. Stopka SA, Rong C, Korte AR, Yadavilli S, Nazarian J, Razunguzwa TT, Morris NJ, Vertes A (2016) Molecular imaging of biological samples on nanophotonic laser desorption ionization platforms. Angew Chem Int Ed 55 (14):4482–4486. https://doi.org/10.1002/anie.201511691
14. Fincher JA, Dyer JE, Korte AR, Yadavilli S, Morris NJ, Vertes A (2018) Matrix-free mass spectrometry imaging of mouse brain tissue sections on silicon nanopost arrays. J Comp Neurol 25(10):24566
15. Walker BN, Antonakos C, Retterer ST, Vertes A (2013) Metabolic differences in microbial cell populations revealed by nanophotonic ionization. Angew Chem Int Ed 52 (13):3650–3653. https://doi.org/10.1002/anie.201207348
16. Walker BN, Stolee JA, Vertes A (2012) Nanophotonic ionization for ultratrace and single-cell analysis by mass spectrometry. Anal Chem 84(18):7756–7762. https://doi.org/10.1021/ac301238k
17. Unal B (2011) Quenching influence of cell culture medium on photoluminescence and morphological structure of porous silicon. Appl Surf Sci 258(1):207–211. https://doi.org/10.1016/j.apsusc.2011.08.032
18. Bhuyan MK, Rodriguez-Devora JI, Fraser K, Tseng TLB (2014) Silicon substrate as a novel cell culture device for myoblast cells. J Biomed

Sci 21. https://doi.org/10.1186/1423-0127-21-47

19. Alvarez SD, Derfus AM, Schwartz MP, Bhatia SN, Sailor MJ (2009) The compatibility of hepatocytes with chemically modified porous silicon with reference to in vitro biosensors. Biomaterials 30(1):26–34. https://doi.org/10.1016/j.biomaterials.2008.09.005
20. Lehto V-P, Vähä-Heikkilä K, Paski J, Salonen J (2005) Use of thermoanalytical methods in quantification of drug load in mesoporous silicon microparticles. J Therm Anal Calorim 80 (2):393–397. https://doi.org/10.1007/s10973-005-0666-x
21. Temiz Y, Lovchik RD, Kaigala GV, Delamarche E (2015) Lab-on-a-chip devices: how to close and plug the lab? Microelectron Eng 132:156–175. https://doi.org/10.1016/j.mee.2014.10.013
22. Coffer JL, Whitehead MA, Nagesha DK, Mukherjee P, Akkaraju G, Totolici M, Saffie RS, Canham LT (2005) Porous silicon-based scaffolds for tissue engineering and other biomedical applications. Phys Status Solidi A 202 (8):1451–1455. https://doi.org/10.1002/pssa.200461134
23. Wang ST, Liu K, Liu JA, Yu ZTF, Xu XW, Zhao LB, Lee T, Lee EK, Reiss J, Lee YK, Chung LWK, Huang JT, Rettig M, Seligson D, Duraiswamy KN, Shen CKF, Tseng HR (2011) Highly efficient capture of circulating tumor cells by using nanostructured silicon substrates with integrated chaotic micromixers. Angew Chem Int Ed 50(13):3084–3088. https://doi.org/10.1002/anie.201005853
24. Fu JP, Wang YK, Yang MT, Desai RA, Yu XA, Liu ZJ, Chen CS (2010) Mechanical regulation of cell function with geometrically modulated elastomeric substrates. Nat Methods 7 (9):733–U795. https://doi.org/10.1038/nmeth.1487
25. De Clercq E (2005) Recent highlights in the development of new antiviral drugs. Curr Opin Microbiol 8(5):552–560. https://doi.org/10.1016/j.mib.2005.08.010
26. Otterbein LE, Bach FH, Alam J, Soares M, Lu HT, Wysk M, Davis RJ, Flavell RA, Choi AMK (2000) Carbon monoxide has anti-inflammatory effects involving the mitogen-activated protein kinase pathway. Nat Med 6 (4):422–428
27. Mirbagheri M, Adibnia V, Hughes BR, Waldman SD, Banquy X, Hwang DK (2019) Advanced cell culture platforms: a growing quest for emulating natural tissues. Mater Horiz 6(1):45–71. https://doi.org/10.1039/c8mh00803e
28. Morris NJ, Anderson H, Thibeault B, Vertes A, Powell MJ, Razunguzwa TT (2015) Laser desorption ionization (LDI) silicon nanopost array chips fabricated using deep UV projection lithography and deep reactive ion etching. RSC Adv 5(88):72051–72057. https://doi.org/10.1039/c5ra11875a

Chapter 12

Compact Quantum Dots for Quantitative Cytology

Phuong Le, Shweta Chitoor, Chunlai Tu, Sung Jun Lim, and Andrew M. Smith

Abstract

In this chapter, we describe the preparation of fluorescent quantum dots for imaging and measuring protein expression in cells. Quantum dots are nanocrystals that have numerous advantages for biomolecular detection compared with organic dyes and fluorescent proteins, but their large size has been a limiting factor. We describe the synthesis of nanoparticles smaller than 10 nm (smaller than an antibody), their attachment to monoclonal antibodies through click chemistry, characterization of the conjugates, and use for labeling of cellular antigens. We further discuss the unique advantages and challenges associated with this approach compared with conventional immunofluorescence techniques.

Key words Immunofluorescence, Fluorescence, Cytology, Single cell, Microscopy, Imaging, Nanocrystal, Nanotechnology, Antibody, Protein, EGFR, Cancer

1 Introduction

Quantitative measurements of nucleic acids and proteins play a central role in clinical disease detection and basic molecular biology. The techniques most commonly employed such as Western blots, enzyme-linked immunosorbent assays (ELISAs), real-time polymerase chain reaction (PCR), and gene chips provide accurate measures of analyte concentrations, but they require liquid specimens [1, 2]. This necessitates cell and tissue homogenization and extraction, which can be a major disadvantage due to destruction of the samples and loss of the native spatial locations of biomolecules within tissues, cells, and subcellular structures. Homogenization also masks rare events within a noisy background and prevents the detection and characterization of cellular subpopulations underlying disease processes [2, 3]. Some of these limitations have been overcome through recent developments in single-cell transcriptomics and single-cell proteomics [4, 5], but these techniques are still challenging and do not retain spatial information.

Bindesh Shrestha (ed.), *Single Cell Metabolism: Methods and Protocols*, Methods in Molecular Biology, vol. 2064, https://doi.org/10.1007/978-1-4939-9831-9_12, © Springer Science+Business Media, LLC, part of Springer Nature 2020

In situ techniques such as immunofluorescent staining and nucleic acid hybridization are widely used to characterize the locations of biomolecules within cells and tissues; however, they suffer from poor quantitative analysis capabilities [6]. In part, this is due to photochemical limitations of organic dyes used as probes to label a bioaffinity molecule (antibody or nucleic acid oligomer) that selectively binds to the analyte of interest. Organic dyes photobleach rapidly, so counting the number of probes in a specific cellular region is very challenging. Moreover, their relatively weak signals often require secondary labeling, for which a polyclonal antibody conjugated to dyes enhances the single binding event of a primary antibody. Unfortunately, the exact number of dyes per antibody and exact number of secondary antibodies per primary antibody are difficult to control and assess. Therefore, correlating the dye signal with a specific number of antigens has been difficult to achieve.

Quantum dots (QDs) are a class of fluorescent nanocrystal that has helped to overcome these limitations due to their unique photophysical characteristics that are well suited for many bio-imaging applications [7–11]. The fluorescence emission from these particles is tens to thousands of times brighter than conventional dyes and highly resistant to photobleaching, so QD signals are intense and stable [12]. Therefore, secondary labeling is usually not necessary and a single antibody-QD labeling step is sufficient. In addition, QDs have broad excitation spectra and narrow emission bands, allowing the simultaneous excitation of many colors that are spectrally distinct, enabling a high degree of multiplexed analyte detection. However, a continual problem has been the large size of QDs. Compared to organic dyes and proteins, which are between 1 and 5 nm in size, conventional quantum dots have a multilayered colloidal structure usually between 15 and 35 nm [13]. This leads to a steric hindrance to binding, slow diffusion, and limited access to antigens in crowded biological environments [13–15].

Recently, a series of new strategies have been advanced to generate QDs with compact hydrodynamic dimensions [15–18]. The key development is the use of thin organic coatings that stabilize the underlying nanocrystal. These coatings are composed of ligands containing multiple binding groups that adsorb to the nanocrystal surface through stable multidentate coordination, in addition to hydrophilic groups that render the nanocrystal dispersible in water. Using recent engineering strategies, these nanocrystals can now be 5–10 nm in diameter, resistant to nonspecific binding, and chemically functionalized for efficient conjugation to biological molecules such as antibodies to allow rapid and specific labeling of cellular antigens. It is hoped that this new generation of QDs will provide the quantitative readout of molecular

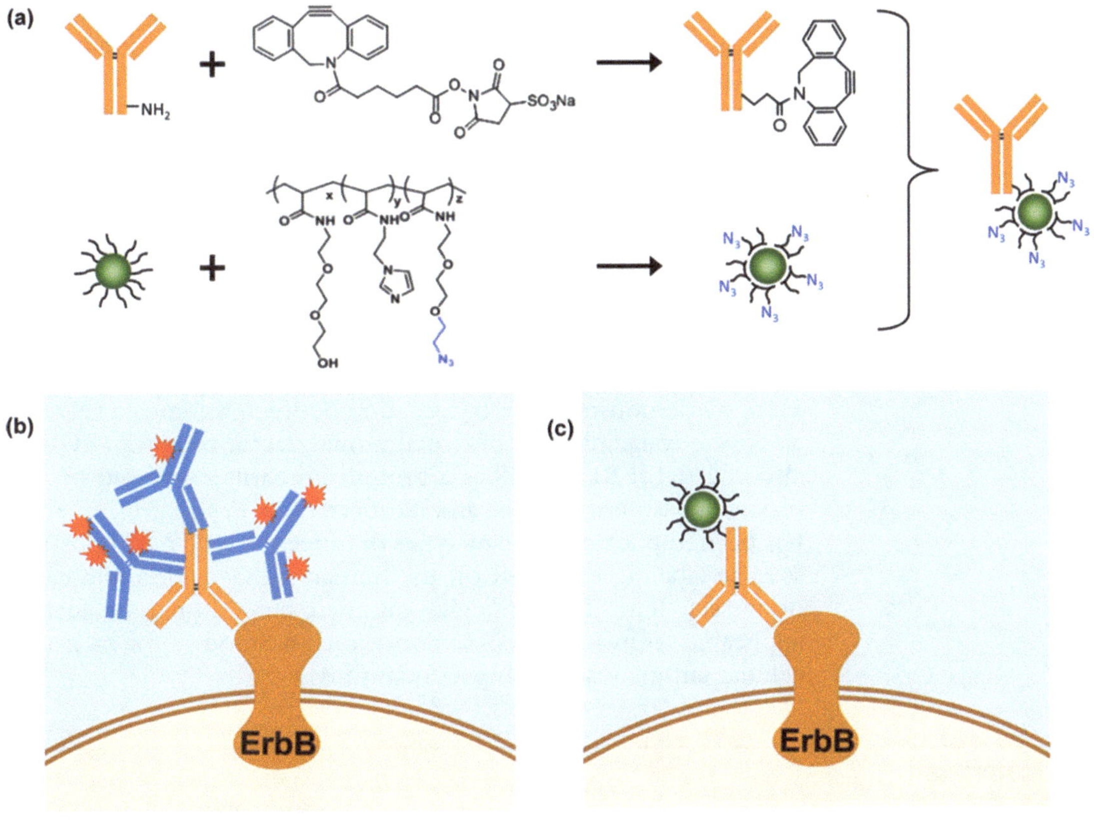

Fig. 1 Schematic depictions of conjugation of compact quantum dots to antibodies using click chemistry and their use for immunofluorescent labeling of cells. (**a**) Primary amines on the lysines or N-termini of antibodies (orange) are mixed with a DBCO-NHS reagent that appends a DBCO group to the antibody. A QD (green) is coated with a multidentate polymer containing azides (blue) that can then be used to conjugate selectively to the DBCO on the antibody. (**b**) Conventional immunofluorescence techniques using secondary labeling of the antigen (here ErbB) involve the binding of a single monoclonal antibody to the antigen, followed by the binding of multiple polyclonal antibodies bound to multiple dyes to enhance the signal. (**c**) With QD-immunofluorescence, only primary labeling is needed due to the bright signal from QDs

concentration in cells and tissues that is currently lacking from conventional immunofluorescence techniques.

However, new challenges arise with the preparation of these probes, which are not yet commercially available. Most importantly, compact QDs and their conjugates are similar in size to proteins, so purification from solutions containing excess macromolecules may not be possible. For this reason, we have developed a methodology to link QDs to antibodies through click chemistry reactions between azides and strained alkynes. These reactions are rapid and efficient in aqueous solution [19]; this eliminates the need for purification and reduces waste of expensive reagents. Figure 1a depicts the general strategy in which an antibody is labeled with the strained alkyne dibenzylcyclooctyne (DBCO) through

N-hydroxysuccinimide (NHS)-mediated coupling [19, 20]. QDs are coated with a multidentate polymer containing imidazole groups that bind to the QD surface [17, 18, 21], and oligoethylene glycol (OEG) groups that provide stability in water and resistance to nonspecific binding [22]. Some of the OEG groups are terminated with azides (N_3) that react with the DBCO group to yield a triazole, covalently linking the QD to the antibody, without the need for catalysts, high temperatures, or reagents that can damage the antibody. It is important to individually optimize all steps involved to ensure the homogeneity and stability of the products. In particular, we find that it is critical to optimize the conjugation between the QD and antibody, which we discuss in this protocol. Here we demonstrate the conjugation between a compact QD and an antibody against the epidermal growth factor receptor (EGFR), also called ErbB1. EGFR is a protein primarily expressed on the plasma membrane of cells, and its aberrant expression or mutation has been implicated in many types of cancer [23]. We demonstrate specific labeling of EGFR on the human A431 epidermoid carcinoma cell line. The strategies outlined here should be further applicable to most forms of monoclonal antibodies for targeting cellular antigens with compact quantum dots.

2 Materials

2.1 Reagents and Buffers

1. Core/shell CdSe/CdZnS quantum dots with a 565 nm emission maximum were synthesized in our lab using standard techniques [24, 25] (*see* **Note 1**).
2. Multidentate azide polymers were synthesized using techniques similar to those reported in the literature [18] (*see* **Note 2**).
3. Monoclonal mouse anti-human EGFR antibody (EGFR.1 clone from BD Science).
4. Control antibody for conjugation optimization (Human IgG Isotype Control, Invitrogen) (*see* **Note 3**).
5. Control antibody for cell labeling (Mouse IgG Isotype Control, Invitrogen) (*see* **Note 3**).
6. Hexane (Sigma-Aldrich).
7. N-methylformamide (NMF) (Sigma-Aldrich).
8. Tetramethylammonium hydroxide (TMAH) 25% methanol solution (Sigma-Alrich).
9. NaOH (Sigma-Alrich).
10. Sterile Phosphate buffer saline (PBS) solution pH 7.4.
11. 1 M Tris–HCl pH 7.4 (Sigma-Alrich).

12. Sodium borate buffer, 500 mM, pH 8.5. Mix anhydrous boric acid (30.92 g) in ultrapure water. Adjust pH to 8.5 using NaOH and dilute to a final volume of 1 L (*see* **Note 4**).
13. Anhydrous DMSO (Sigma-Alrich).
14. DBCO-sulfo-NHS ester (Click Chemistry Tools).
15. DMEM with sodium pyruvate (ATCC Catalog No. 30-2002).
16. Trypsin-EDTA (Life Technologies).
17. Nunc Lab-Tek II Chambered Coverglass (Thermo Scientific).
18. Hoechst stain (Invitrogen).
19. 2-Azidoacetic acid 97% (Sigma-Alrich).
20. Ammonium persulfate (APS) (Sigma-Aldrich).
21. TEMED (Sigma-Alrich).
22. 32% Paraformaldehyde (PFA, Electron Microscopy Science).
23. Triton X-100 (Sigma-Alrich).
24. 1% (w/v) Bovine serum albumin (BSA) (GE Healthcare): dissolve 500 mg of BSA powder in 50 mL PBS.
25. Fetal bovine serum (FBS) (VWR).
26. Acrylamide/*N*,*N*′-Methylenebisacrylamide (19:1, 40% mix solution in water) (Acros Organics).
27. Agarose (Fisher Scientific).
28. Glycerol (Sigma-Alrich).
29. 6× gel loading buffer: 30% (w/v) glycerol in 50 mM sodium borate. Store at room temperature.

2.2 Special Equipment

1. Spectra/Por 7 dialysis tubing with molecular weight cutoff (MWCO) of 50,000 Da (Spectrum Labs).
2. Amicon Ultra Centrifugal Filter with MWCO of 3000 Da (0.5 mL) or 50,000 Da (15 mL) (Millipore).
3. Eppendorf centrifuge 5424 R.
4. NanoDrop ND-1000 Spectrophotometer.
5. Molecular Imager Gel Doc XR system (Bio-rad).
6. Horizontal Mini-Gel Electrophoresis apparatus with 7.5 cm × 10 cm ($w \times l$) gel bed (model MGU-202T, CBS Scientific).
7. Zeiss Axio Observer Z1 inverted microscope with 100× 1.45 NA alpha Plan-Fluar oil immersion objective.
8. Hand-held long wave UV lamp.
9. Cary Series 5000 UV-Vis-NIR Spectrophotometer (Agilent).

3 Methods

3.1 Polymer Coating of QDs with Click-Reactive Azides

1. Procedures for phase transfer should be performed using standard air-free techniques to minimize exposure of the nanoparticles and chemicals involved to air and water.
2. Disperse purified CdSe/CdZnS QDs in hexane (0.2 mL, 2 nmol) and mix with NMF (0.2 mL) to form a 2-phase mixture in a glass bottle.
3. While stirring, add tetramethylammonium hydroxide (TMAH) 25% methanol solution (52 μL) to transfer the QDs from hexane to NMF. Stir vigorously for 1 h.
4. Remove the hexane phase and wash the NMF dispersion by twice stirring with an equal volume of pure hexane for 5 min. Evaporate the residual hexane under vacuum.
5. Centrifuge the QD suspension at 7000 × *g* for 10 min to remove possible aggregates, and then dilute with anhydrous DMSO (1 mL).
6. Add a DMSO solution of the multidentate polymer in excess (3.6 mg in 0.1 mL DMSO) to the QDs while stirring.
7. Bubble the solution with nitrogen gas for 5 min and heat to 110 °C while stirring for 2 h.
8. Cool the solution to room temperature and add a dilute aqueous solution of NaOH (1 mM, 1 mL). Stir for 10 min.
9. Load the solution into a dialysis bag and dialyze against NaOH (1 mM) aqueous solution for 1 h with four changes of the solvent to remove DMSO.
10. Purify the QDs using a centrifugal filter (50,000 Da MWCO) and exchange buffers with sodium borate (50 mM), following instructions provided by the manufacturer.
11. Centrifuge the QDs at 7000 × *g* for 10 min to remove possible aggregates at store at 4 °C until use.

3.2 Conjugation of Antibody to Cyclooctyne

1. Prepare antibody (IgG) in PBS that does not contain azides (*see* **Note 5**).
2. Immediately before conjugation, prepare a 10 mM DBCO-sulfo-NHS ester solution by dissolving 1 mg of DBCO-sulfo-NHS ester powder in 188 μL of anhydrous DMSO in a glass vial then vortexing to dissolve the powder (*see* **Note 6**).
3. Mix the DBCO-sulfo-NHS solution with the IgG solution. Use a DBCO:IgG molar ratio of 5 if the IgG concentration is equal to or greater than 5 mg/mL; otherwise use a molar ratio of 50.
4. Incubate the reaction for 2 h on ice.

5. Quench the reaction by adding Tris–HCl to a final concentration of 50 mM.
6. Purify the IgG_{DBCO} conjugate using centrifugal filters (3000 Da MWCO) according to the instructions provided by the manufacturer.
7. Confirm IgG conjugation to DBCO by measuring the ultraviolet absorption spectrum, and determine the IgG concentration according to the instructions provided by the manufacturer. For quantification purposes, the extinction coefficient of the DBCO functional group at 309 nm is $\varepsilon_{309} = 12{,}000\ M^{-1}\ cm^{-1}$.

3.3 Conjugation Between Cyclooctyne-Activated Antibody and QD-Azide

1. Exchange the buffer for QD-azide from 50 mM sodium borate to PBS by using a centrifugal filter (3000 Da MWCO), following instructions provided by the manufacturer.
2. Centrifuge the QDs at 7000 × g for 10 min and retain the supernatant in order to remove any aggregates.
3. Measure the QD concentration using a NanoDrop Spectrophotometer. Typically, 2 μL of the solution is used and the QD concentration can be calculated from the known extinction coefficient.
4. Mix the QDs with IgG_{DBCO} at a ratio optimal for the QDs. It is important to optimize both the ratio and the time for complete conjugation in order to minimize the number of non-reacted antibodies and to prevent aggregation of the products. In the data shown in Fig. 2, we find that the reactions are complete in ~7 h at 4 °C and that conjugations are ideal for a IgG:QD ratio less than or equal to 2.5. We typically use a molar excess of 0.6–2.5 to ensure that the majority of QDs have just one or two IgGs. Pipet up and down several times to mix and incubate at the optimum time at 4 °C.
5. Quench unreacted DBCO groups by adding a 50-fold molar excess of 2-azidoacetic acid for 15 min on ice. Prepare a 13 mM 2-azidoacetic acid solution by diluting the 2-azidoacetic acid stock 1000-fold. Then add 1.65 μL of 13 mM 2-azidoacetic acid into 10 μL of 10 μM of QD-IgG solution; scale as needed.
6. Store the QD-IgG conjugates at 4 °C and use within 3 days after conjugation (long-term stability is currently unknown). Periodically check for aggregates in the QD stock. Centrifuge 7000 × g for 10 min to remove possible aggregates before use.

3.4 Gel Electrophoresis to Analyze Antibody Conjugated QDs (QD-IgG)

1. Analyze QD-IgG conjugates using electrophoresis in hybrid polyacrylamide–agarose gels (2% polyacrylamide and 0.5% agarose), which have previously been demonstrated to yield improved resolution compared to conventional gels [26].

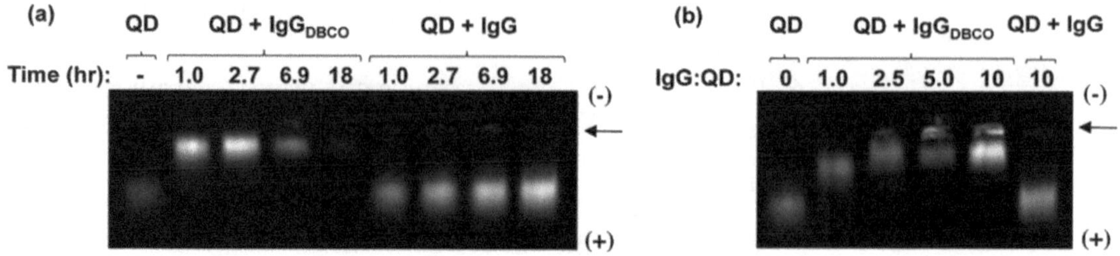

Fig. 2 Optimization of QD-antibody conjugation. QDs were mixed with IgG conjugated to DBCO (IgG_{DBCO}) or IgG alone for the specified times and at the specified molar ratios. (**a**) QDs were mixed with IgG at an IgG:QD ratio of 5:1 and allowed to react for the specified times. The first lane shows the QDs alone. With conjugation to IgG_{DBCO}, the migration distance is reduced, indicating successful conjugation. The migration distance does not change when IgG alone is used. (**b**) QDs were mixed with IgG at the indicated IgG:QD ratios and allowed to react for 7 h. The first lane shows the QDs alone. With increasing amounts of IgG_{DBCO}, the migration distance is reduced, indicating successful conjugation. The original QD band is not present for the 1.0 ratio, and migration shifts cease after the 2.5 ratio. A greater amount of aggregated conjugate is observed in the well for the 5.0 and 10 ratios. The migration distance does not change when IgG alone is used. The electrophoresis electrodes are indicated by the (−) and (+) notations; all of the samples migrated toward the positive pole, indicating a net negative electrostatic charge in the pH 8.5 buffer. The wells are indicated with arrows

2. Prepare 25 mL of 1% (w/v) melted agarose in water and 25 mL of 4% (v/v) acrylamide in sodium borate buffer (100 mM).
3. Mix the two solutions in a 250 mL flask and incubate in a 55 °C water bath for 2 min.
4. Slowly add ammonium persulfate and TEMED to final concentrations of 0.05% (w/v) and 1:2500 (v/v), respectively, and immediately pour the solution into the gel apparatus.
5. Allow the gel to solidify for 1 h.
6. Prepare 250 mL of 50 mM sodium borate buffer as running buffer. Pour the solution over the gel so that it is completely immersed.
7. Load samples: mix 10 μL of each sample with 2 μL of the loading buffer and then load into the wells.
8. Run the conjugate samples in adjacent wells to QDs for which no IgG was added, as well as QDs for which IgG was added, but without conjugation to DBCO.
9. Run the electrophoresis at 120 V for 20–30 min at 4 °C.
10. Image the gel using a Molecular Imager Gel Doc™ XR System (Bio-Rad) in Trans UV mode.
11. Representative gel images are shown in Fig. 2.

3.5 Staining A431 Cells with QD-IgG Conjugates Targeting Human EGFR

1. Trypsinize A431 cells from a stock culture and transfer 50,000 cells in 500 μL complete medium (DMEM with 10% FBS) to each well of a Lab-Tek glass-bottom 8-well chamber and allow cells to adhere for 12–24 h in a CO_2 incubator at 37 °C (*see* **Note 7**).

2. Aspirate media and gently rinse cells twice with PBS.
3. Prepare 4% PFA solution by diluting 32% PFA stock solution in PBS (*see* **Note 8**).
4. Add 300 μL of 4% PFA solution to each well and incubate for 20 min at room temperature.
5. Aspirate PFA solution and rinse three times with PBS. Each time, allow 5 min before aspirating.
6. Prepare 0.1% (v/v) Triton X-100 in PBS by dissolving 25 μL of Triton X-100 in 25 mL of PBS, and vortex to mix if necessary.
7. Add 300 μL of 0.1% (v/v) Triton X-100 to each well and incubate for 20 min at room temperature.
8. Aspirate the Triton X-100 solution and rinse 3 times with PBS.
9. Add 1% BSA with 5% FBS to each well and incubate for 1 h at room temperature to block.
10. Prepare 10 nM of QD-Ab solutions by diluting QD-Ab in 1% BSA.
11. Aspirate the blocking media and add 100 μL of 10 nM QD-Ab targeting EGFR to each well and incubate for 30 min at room temperature. Control groups should be included by replacing some QD-Ab solutions with solutions of QDs bound to control antibodies, solutions of QDs not conjugated to antibodies, or solutions with no QDs. Optionally, these controls can be performed on cells known to not express the EGFR antigen, or the antigen can be pre-blocked with free antibody.
12. Rinse twice with 1% BSA and once with PBS.
13. Stain nuclei with 1:5000 (v/v) of Hoechst in 1% BSA for 10 min at room temperature. Prepare the Hoechst staining solution by dissolving 0.2 μL of Hoechst stock in 1 mL of 1% BSA solution. From this point onward, work in the dark to prevent photobleaching the nuclear dye.
14. Rinse twice with 1% BSA and twice with PBS.
15. Add 300 μL of PBS to each well and image immediately.

3.6 Fluorescence Microscopy and Image Analysis

1. Examine the cell sample using transillumination imaging mode at 100× and select a region of interest (ROI) containing cells.
2. Bring cells into focus using the blue Hoechst channel (*see* **Note 9**).
3. For the same ROI, maintain the same *z*-axis and collect the QD emission in epifluorescence mode using 405 or 488 nm laser excitation and an emission filter appropriate for the particular QDs (*see* **Note 10**).
4. Adjust brightness and contrast of the fluorescence images using Image-J (*see* **Note 11**).
5. Example images are provided in Fig. 3.

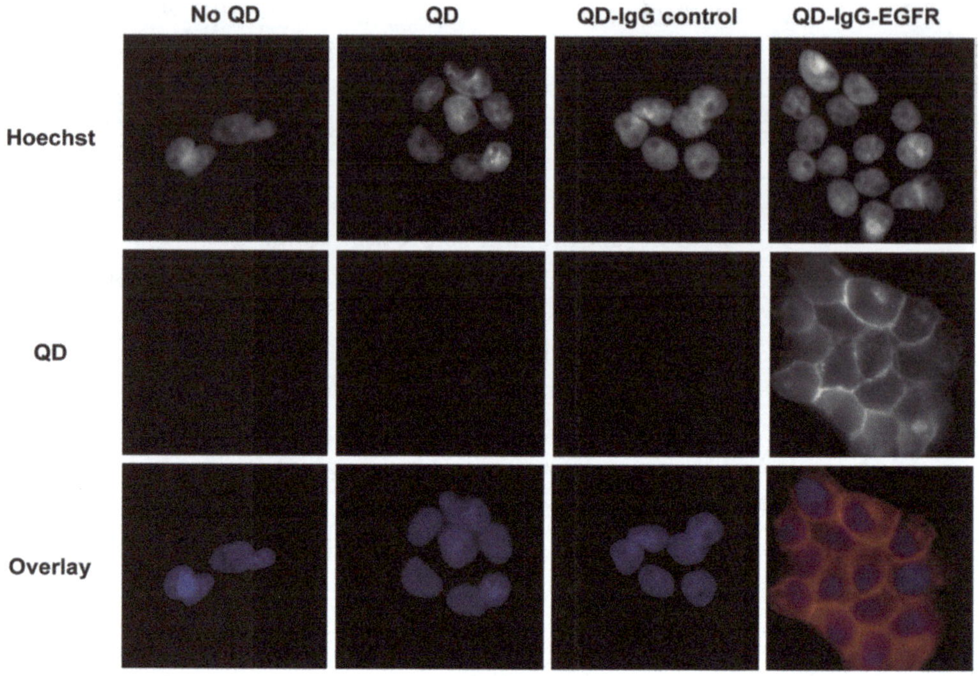

Fig. 3 Typical images obtained for QD staining of the EGFR antigen in A431 cells. Different rows depict (top) the blue fluorescence channel for the Hoechst stain (nuclei), (middle) the yellow channel for the QD stain (EGFR antigen), and (bottom) an overlay of the two for which the Hoechst channel is false-colored blue and the QD channel is false-colored red. From left to right, the columns correspond to samples for which (1) no QD was added, (2) QDs alone were added without conjugation to an antibody, (3) QDs conjugated to the control antibody were added, and (4) QDs conjugated to the EGFR-antibody were added. The only channel with substantial QD staining was column 4

4 Notes

1. During QD synthesis, it is critical to precisely determine the extinction coefficient using reported values for the CdSe core [27] and carefully monitor changes during CdZnS shell growth. It is optimal to measure the values in the 350–450 nm range where there are few discrete absorption peaks. In our case, the extinction coefficient at 350 nm was $\varepsilon_{350} = 1{,}400{,}816\ \text{cm}^{-1}\ \text{M}^{-1}$ and the extinction coefficient at the first exciton peak at 544 nm was $\varepsilon_{544} = 122{,}357\ \text{cm}^{-1}\ \text{M}^{-1}$.
2. A variety of polymers that bind to QDs through multidentate coordination have been reported in the literature based on a polymeric backbone like polyacrylic acid or polymaleic anhydride [16, 17, 21, 28]. For the purposes of this work, it is important that the polymer has (a) multiple QD-binding groups, such as thiols or imidazoles, (b) azides for click chemistry-based conjugations, and (c) a large number of OEG or zwitterionic functional groups to minimize nonspecific interactions with cells. The polymer that we used was

reported previously by our group [18], with 35% of the side chains as imidazoles, 45% as OEG-hydroxyls, and 20% as OEG-azides.

3. We recommend optimizing and practising antibody conjugations using an inexpensive IgG such as the isotype control antibody, as monoclonal antibodies are often expensive. The results translate well between different IgG antibodies.
4. Unless specified otherwise, all reagents are of the best analytical grade and all solutions should be prepared in Milli Q-water (resistivity of 18.2 MΩ/cm at 25 °C).
5. Many commercial primary antibodies are provided with BSA or gelatin, which will interfere with the conjugation process. It is best to purchase antibodies that contain no additional proteins. In addition, commercial antibodies often contain sodium azide as an antibacterial agent, which should be removed via dialysis before conjugation due to reactivity with cyclooctynes.
6. NHS groups can be rapidly inactivated when exposed to moisture and aqueous solutions at pH higher than 6. It is best to handle DBCO-sulfo-NHS ester powder under a nitrogen environment.
7. Cells can also be cultured on glass coverslips.
8. The formaldehyde fixation solution should be prepared fresh. Handle the formaldehyde solution in a fume hood only. Wear appropriate gloves and dispose of formaldehyde waste according to procedures appropriate for the lab.
9. The filter cube used for Hoechst has a 365 nm excitation filter and 445/50 nm emission filter.
10. The QD filter should be selected to overlap maximally with the QD emission spectrum. For our QD samples, we used a 562/40 bandpass emission filter. Exposure time and gain can be adjusted accordingly to achieve a signal that spans the range of the detector without saturation. All samples including controls should be imaged using the same settings. Therefore, it is best to image the sample with highest QDs fluorescence signal first.
11. Brightness and contrast adjustment should be performed identically for all samples.

References

1. Liu R, Wu P, Yang L et al (2014) Inductively coupled plasma mass spectrometry-based immunoassay: a review. Mass Spectrom Rev 33:373–393
2. Lu J, Tsourkas A (2009) Imaging individual microRNAs in single mammalian cells in situ. Nucleic Acids Res 37:e100
3. Xing Y, Smith AM, Agrawal A et al (2006) Molecular profiling of single cancer cells and clinical tissue specimens with semiconductor quantum dots. Int J Nanomedicine 1:473–481
4. Zong C, Lu S, Chapman AR et al (2012) Genome-wide detection of single-nucleotide

and copy-number variations of a single human cell. Science 338:1622–1626

5. Xue Z, Huang K, Cai C et al (2013) Genetic programs in human and mouse early embryos revealed by single-cell RNA sequencing. Nature 500:593–597
6. Chen C, Peng J, Xia H-S et al (2009) Quantum dots-based immunofluorescence technology for the quantitative determination of HER2 expression in breast cancer. Biomaterials 30:2912–2918
7. Zrazhevskiy P, Gao X (2013) Quantum dot imaging platform for single-cell molecular profiling. Nat Commun 4:1619
8. Xu H, Xu J, Wang X et al (2013) Quantum dot-based, quantitative, and multiplexed assay for tissue staining. ACS Appl Mater Interfaces 5:2901–2907
9. Kairdolf BA, Smith AM, Stokes TH et al (2013) Semiconductor quantum dots for bioimaging and biodiagnostic applications. Annu Rev Anal Chem 6:143–162
10. Le P, Lim SJ, Baculis BC, Chung HJ, Kilian KA, Smith AM (2019) Counting growth factors in single cells with infrared quantum dots to measure discrete stimulation distributions. Nat Commun 10(909). (2019) https://www.nature.com/articles/s41467-019-08754-5
11. Liu Y, Le P, Lim SJ, Ma L, Sarkar S, Han Z, Murphy SJ, Kosari F, Vasmatzis G, Cheville JC, Smith AM (2018) Enhanced mRNA FISH with compact quantum dots. Nat Commun 9 (4461). https://www.nature.com/articles/s41467-018-06740-x
12. Chan WCW, Maxwell DJ, Gao X et al (2002) Luminescent quantum dots for multiplexed biological detection and imaging. Curr Opin Biotechnol 13:40–46
13. Smith AM, Nie SM (2009) Next-generation quantum dots. Nat Biotechnol 27:732–733
14. Pinaud F, Clarke S, Sittner A et al (2010) Probing cellular events, one quantum dot at a time. Nat Methods 7:275–285
15. Howarth M, Liu W, Puthenveetil S et al (2008) Monovalent, reduced-size quantum dots for imaging receptors on living cells. Nat Methods 5:397–399
16. Smith AM, Nie S (2008) Minimizing the hydrodynamic size of quantum dots with multifunctional multidentate polymer ligands. J Am Chem Soc 130:11278–11279
17. Liu W, Greytak AB, Lee J et al (2010) Compact biocompatible quantum dots via RAFT-mediated synthesis of imidazole-based random copolymer ligand. J Am Chem Soc 132 (2):472–483
18. Ma L, Tu C, Le P, Chitoor S, Lim SJ, Zahid MU, Teng KW, Ge P, Selvin PR, Smith AM (2016) Multidentate polymer coatings for compact and homogeneous quantum dots with efficient bioconjugation. J Am Chem Soc 138(10):3382–;3394. https://pubs.acs.org/doi/10.1021/jacs.5b12378
19. Baskin JM, Prescher JA, Laughlin ST et al (2007) Copper-free click chemistry for dynamic in vivo imaging. Proc Natl Acad Sci 104:16793–16797
20. Thorek DLJ, Elias DR, Tsourkas A (2009) Comparative analysis of nanoparticle-antibody conjugations: carbodiimide versus click chemistry. Mol Imaging 8:221–229
21. Zhang P, Liu S, Gao D et al (2012) Click-functionalized compact quantum dots protected by multidentate-imidazole ligands: conjugation-ready nanotags for living-virus labeling and imaging. J Am Chem Soc 134:8388–8391
22. Zheng M, Davidson F, Huang X (2003) Ethylene glycol monolayer protected nanoparticles for eliminating nonspecific binding with biological molecules. J Am Chem Soc 125:7790–7791
23. Yarden Y (2001) The EGFR family and its ligands in human cancer: signalling mechanisms and therapeutic opportunities. Eur J Cancer 37:3–8
24. Smith AM, Nie S (2012) Compact quantum dots for single-molecule imaging. J Vis Exp 68: e4236
25. Xie RG, Kolb U, Li JX et al (2005) Synthesis and characterization of highly luminescent CdSe-Core CdS/$Zn_{0.5}Cd_{0.5}S$/ZnS multishell nanocrystals. J Am Chem Soc 127 (20):7480–7488
26. Liu HY, Gao X (2011) Engineering monovalent quantum dot–antibody bioconjugates with a hybrid gel system. Bioconjug Chem 22:510–517
27. Jasieniak J, Smith L, Jv E et al (2009) Re-examination of the size-dependent absorption properties of CdSe quantum dots. J Phys Chem C 113:19468–19474
28. Palui G, Na HB, Mattoussi H (2012) Poly (ethylene glycol)-based multidentate oligomers for biocompatible semiconductor and gold nanocrystals. Langmuir 28:2761–2772

Chapter 13

Ambient Lipidomic Analysis of Single Mammalian Oocytes and Preimplantation Embryos Using Desorption Electrospray Ionization (DESI) Mass Spectrometry

Christina R. Ferreira, Valentina Pirro, Alan K. Jarmusch, Clint M. Alfaro, and R. Graham Cooks

Abstract

Desorption electrospray ionization (DESI) is a spray-based ambient ionization method for mass spectrometry (MS) that generates ions in native atmospheric conditions (e.g., pressure and temperature). Ambient ionization allows in situ analysis of unmodified samples by eliminating analyte extraction and separation steps before MS. Lipid analysis of individual mammalian oocytes and preimplantation embryos is challenging because of their complex chemical composition and minute dimensions (100–300 μm in diameter). Nonetheless, DESI-MS can provide comprehensive lipidomic profiles of individual oocytes or embryos using a fast and simple workflow. DESI-MS lipid profiles include many classes of lipids such as phosphatidylcholines (PC), triacylglycerols (TAG), free fatty acids (FFA), phosphatidylethanolamines (PE), phosphatidylinositols (PI), phosphatidylserines (PS), diacylglycerols (DAG), ubiquinone, cholesterol and cholesterol derivatives (e.g., cholesterol sulfate and cholesterol esters). Depending on the mass spectrometer used, there is the additional possibility of obtaining structural information of lipids via MS^n, and molecular formulae via high resolution MS. Here, we describe the procedures for sample handling, the DESI-MS experimental arrangement, and the data collection and data analysis using low and high-resolution mass spectrometers (such as a linear ion trap and Orbitrap mass analyzer, respectively), as well as strategies for structural characterization of lipids. Lastly, we detail our workflow for relating the chemical information obtained by DESI-MS into the biological context of oocyte and embryo metabolism.

Key words DESI-MS, Reactive DESI, Ambient mass spectrometry, Single cell, Multidimensional MS scan, Exact mass measurement, Multivariate statistics, Lipids

1 Introduction

Ambient ionization refers to the generation of ions under ambient conditions (e.g., native temperature and pressure), requiring little-to-no sample preparation [1–3]. Ambient techniques coupled to mass spectrometry (MS) were first described a little over a decade ago and have since expanded into a large family with dozens of individual methods [3, 4]. The ability to examine a sample in its

Bindesh Shrestha (ed.), *Single Cell Metabolism: Methods and Protocols*, Methods in Molecular Biology, vol. 2064, https://doi.org/10.1007/978-1-4939-9831-9_13,

native state (thus eliminating extraction and separation procedures) and speed are the primary advantages of ambient ionization [5].

Desorption electrospray ionization (DESI) was the first ambient ionization technique reported [3]. DESI has been applied in over 500 scientific publications (Web of Science, July 2015) for the qualitative and quantitative analysis of a large range of molecules, including metabolites [6], drugs [7], lipids [8], and chemical agents [9]. A few years ago, developments in both DESI source design and sample analysis protocols, such as the introduction of histologically compatible solvents [10, 11], paved the way for application in developmental biology, particularly the chemical profiling of individual mammalian oocytes and preimplantation embryos. DESI-MS profiling of lipids, particularly lipids that compose the cell membrane and are involved in cellular signaling, obtained from the analysis of tissue has been the focus of several studies, in which it was observed that the lipid profile is characteristic of different tissue types, disease states (e.g., normal vs. tumor), and metabolism [12].

DESI-MS analysis of mammalian oocytes or preimplantation embryos is performed by placing a sample onto a glass microscope slide and spraying with charged droplets formed with the aid of a N_2 gas stream directed at the sample. As the DESI spray of charged droplets, i.e., primary droplets, impacts the sample surface, a thin film of solvent is created, allowing for analyte dissolution. Later arriving primary droplets sputter smaller secondary droplets containing analytes from the surface where they undergo evaporation and Coulombic fission, generating ions via electrospray-like processes. Physicochemical properties of the small molecules present in these complex samples dictate the conditions to be used for DESI-MS analysis, such as the use of positive or negative ionization mode and the choice of a DESI solvent [13].

The introduction of nondestructive (or histologically compatible) solvent systems allowed the analysis of individual mammalian oocytes and preimplantation embryos by DESI-MS [10]. Histologically compatible solvents are specific solvent combinations of dimethylformamide (DMF) with acetonitrile (ACN), ethanol, methanol, and other organic solvents that allow chemical information to be obtained while preserving sample morphology. Specifically for mammalian oocytes and preimplantation embryos, histologically compatible solvents such as DMF-ACN, 1:1 (v/v), and pure ACN doped with <5 ppm silver nitrate ($AgNO_3$) can be used to gently extract the lipids from oocytes and embryos while providing signal for many seconds so that mass spectra can be accumulated and averaged to generate a representative lipid profile [14, 15].

Here, we describe and comment in detail the procedure of performing in situ lipidomic analysis of individual oocytes and preimplantation embryos from species such as mouse, bovine, and swine by DESI-MS. Reagents, MS instrumentation, and DESI conditions are listed. Classes of lipids detected using specific solvent systems are described as well as procedures for performing lipid attribution using MS^n and high-resolution MS (HRMS). We also illustrate the statistical workflow and data analysis for interpreting chemical information and the implications in developmental biology. The methods of lipid analysis used here are widely applicable even though embryological applications are emphasized.

2 Materials

Reagents should be of least analytical grade (ideally MS grade) and prepared daily at room temperature unless otherwise stated. Chemicals were used as received without further purification. Please follow chemical and biohazard waste disposal regulations after sample analysis.

2.1 Samples

1. Fresh or frozen samples can be used (*see* **Note 1**).
2. Fresh mammalian immature oocytes isolated from ovaries; mature oocytes and preimplantation embryos in culture media (*see* **Note 2**).
3. Frozen mammalian oocytes and preimplantation embryos stored in a minimum volume ($\leq$5 μL) of phosphate buffer saline supplemented with 1% polyvinyl alcohol (PVA) in siliconized cryovials and stored at −80 °C. If sample transportation is needed, it can be performed using dry ice (*see* **Note 3**).
4. Frozen mouse brain tissue cryosectioned (typically 15 μm of thickness) and thaw-mounted onto glass microscope slides serves as reference sample for ion signal and DESI spray optimization.

2.2 Materials and Reagents

1. Fresh or frozen oocytes and embryos (*see* Table 1 for reported species of oocyte and embryo samples analyzed by DESI-MS).
2. Appropriate pipetting system (typically glass capillary) to handle oocytes and embryos (*see* **Note 4**).
3. Siliconized microtubes (*see* **Note 5**).
4. Glass microscope slides (Gold Seal, Portsmouth, NH, USA) (*see* **Note 6**).
5. Black marker (*see* **Note** 7).
6. Phosphate buffer saline (PBS).
7. Polyvinyl alcohol (PVA).
8. Acetonitrile (ACN).
9. Methanol (MeOH).

Table 1
Lipid classes detected, species for which oocytes and preimplantation embryos have been analyzed by DESI-MS, and ionization modes

Class	Biological sample	Ionization mode	References
Free fatty acids/free fatty acid dimers	Porcine oocytes, mouse oocytes and embryos, bovine embryos	(−)	[14, 15, 17, 26, 27]
Phosphatidylcholine (PC)	Bovine oocytes and embryos	(−)	[14, 15, 17, 27]
Phosphatidylinositol (PI)	Porcine oocytes, mouse and bovine oocytes and embryos	(−)	[14, 15, 17, 26, 27]
Phosphatidylserine (PS)	Porcine oocytes, mouse and bovine oocytes and embryos	(−)	[14, 26, 27]
Phosphatidylethanolamine (PE)	Mouse and bovine oocytes and embryos	(−)	[14, 15, 17, 27]
Squalene, cholesterol esters and ubiquinone	Porcine oocytes, bovine oocytes and embryos	(+) with $AgNO_3$	[14, 26, 27]
Cholesterol sulfate	Bovine oocytes and embryos	(−)	[14, 27]
Diacylglycerol (DAG)	Porcine oocytes	(+) with $AgNO_3$	[26]
Tryacylglycerol (TAG)	Porcine oocytes, bovine oocytes and embryos	(+) with $AgNO_3$	[14, 26, 27]
Phosphatidylglycerol (PG)	Porcine oocytes, bovine oocytes and embryos	(−)	[14, 15, 17, 26, 27]

10. Nanopure water (18.2 MΩ cm^{-1}).
11. Dimethylformamide (DMF).
12. Silver nitrate (*see* **Note 8**) ($AgNO_3$).

2.3 Mass Spectrometry Instrumentation

1. DESI source and moving stage (*see* **Note 9**).
2. Mass spectrometer (Orbitrap or linear ion trap mass spectrometer, Thermo Fisher Scientific, San Jose, CA) (*see* **Note 10**). The only modifications to the instruments were the installation of a custom DESI source, which included a source override adapter, an external cable for the application of the high voltage, and an extended stainless steel ion transfer capillary (total length, 180 mm, inner diameter, 0.02″, outer diameter, 1/16″, length protruding from the MS vacuum system, ~87.5 mm). Swagelok stainless steel fittings were used to connect the capillary with the threaded MS inlet. The 1/16" Swagelok fittings were machined so that the capillary can be placed all the way through the fitting and secured on one side with a ferrule and metal nut.

2.4 Other Instrumentation

1. Stereomicroscope.
2. Ultra-low temperature freezer (−80 °C).
3. Electronic desiccator (*see* **Note 11**).
4. Precision syringe pump and compatible glass syringes (*see* **Note 12**).

3 Methods

3.1 Sample Handling

1. Prepare sample washing solution 1 (SWS 1): phosphate buffer saline supplemented with 1% PVA.
2. Prepare sample washing solution 2 (SWS 2): Methanol (MeOH): nanopure water (1:1, v/v) (*see* **Note 13**).
3. Draw small circles on the back of the glass microscope slides (the samples will be placed in each circle) with the black marker. Number the circles and add identify label to the glass microscope slide (*see* **Note 14**).
4. Using a stereomicroscope, transfer samples from the culture medium to droplets of ~100 μL of SWS 1. At least three washing steps are recommended (*see* **Note 15**).
5. Using a stereomicroscope, transfer oocytes and/or embryos from SWS 1 to SWS 2 (also droplets of ~100 μL). This washing procedure should also be performed three times and should be fast (*see* **Note 16**).
6. Using a stereomicroscope, individually transfer samples from SWS2 to the glass slides and place each sample inside one circle drawn on the back of the microscope slide. A small amount of SWS 2 (around 2 μL) should be placed with the sample onto the glass slide (*see* **Note 17**).
7. Let the residue of SWS 2 dry at room temperature on the glass slide or place it in an electric desiccator for a few minutes until the liquid around the samples is evaporated.
8. Place the glass slides with samples in a slide box and store the box at −80 °C until the time of analysis.

3.2 DESI-MS Reagents Preparation

1. DESI spray solution for negative ion mode ionization: Dimethylformamide (DMF)—acetonitrile (ACN) (1:1, *v/v*) (*see* **Note 18**).
2. DESI spray solution for positive ion mode ionization: Acetonitrile with 3.5–5 ppm silver nitrate (*see* **Note 19**).

3.3 DESI-MS Optimization

1. Optimize the DESI spray to obtain a small and stable spray spot. (*see* Eberlin 2014 [16] for additional details on how to build and optimize a DESI spray) (*see* Fig. 6 for a tutorial on how to build a DESI spray). Direct the DESI spray onto the

top of a glass slide. The bottom of the glass slide should be made opaque with a black marker, which aids in visualizing the spray spot and examining its stability.

2. Optimize the flow rate of the DESI spray solvent system and pressure of the nitrogen gas. Negative ionization mode using DMF-ACN (1:1) has a flow rate usually between 0.7 and 1 μL min^{-1}. Positive ionization mode using ACN doped with $AgNO_3$, the flow rate is usually between 6 and 8 μL min^{-1}. Nitrogen gas is commonly set to values between 150 and 170 psi.
3. Optimize the DESI spray geometry using mouse brain tissue [16] and then with a test samples of oocytes or preimplantation embryos. Common conditions for DESI spray geometry are: incident angle, 52°; tip-to-source distance, about 2 mm; tip-to-inlet distance, about 4 mm. Optimize the geometry further by slightly changing these parameters in order to obtain the maximum and more stable signal intensity, and a stable fine sampling spot.
4. Tuning of the mass spectrometer using an analytical standard, ideally the molecule of interest or a similar mass-to-charge (*m/z*) ratio, can improve sensitivity. This can be done using brain tissue. The tuning procedure optimizes capillary voltage, tube lenses, quadrupoles, and hexapoles (depending on the instrument) potentials to maximize ion transmission. Tuning can also be done with an alternative ionization technique, such as nano-electrospray ionization.
5. Optimize capillary temperature and high voltage applied to the stainless steel syringe pump. Common values are 275 °C and 5 kV, respectively, for both positive and negative ion mode acquisition.
6. Optimize injection time and the number of microscans to produce a stable and high-quality MS signal intensity throughout the mass range of interest, without space charging (deviation in ion trajectories that impacts ion *m/z* values and resolution in the mass spectra. This phenomenon is most noticeable in ion trap instruments that confine large numbers of ions in a small volume). The automatic gain control (AGC) can be enabled in order to compensate for variance in ion abundance.

3.4 DESI-MS Analysis

1. An overview of the whole experimental workflow of DESI-MS analysis of mammalian oocytes and embryos is illustrated in Fig. 1.
2. For the analysis of frozen samples, remove the glass slide containing the samples from the freezer and place it in the vacuum desiccator for a few minutes, until all visible water has been removed.

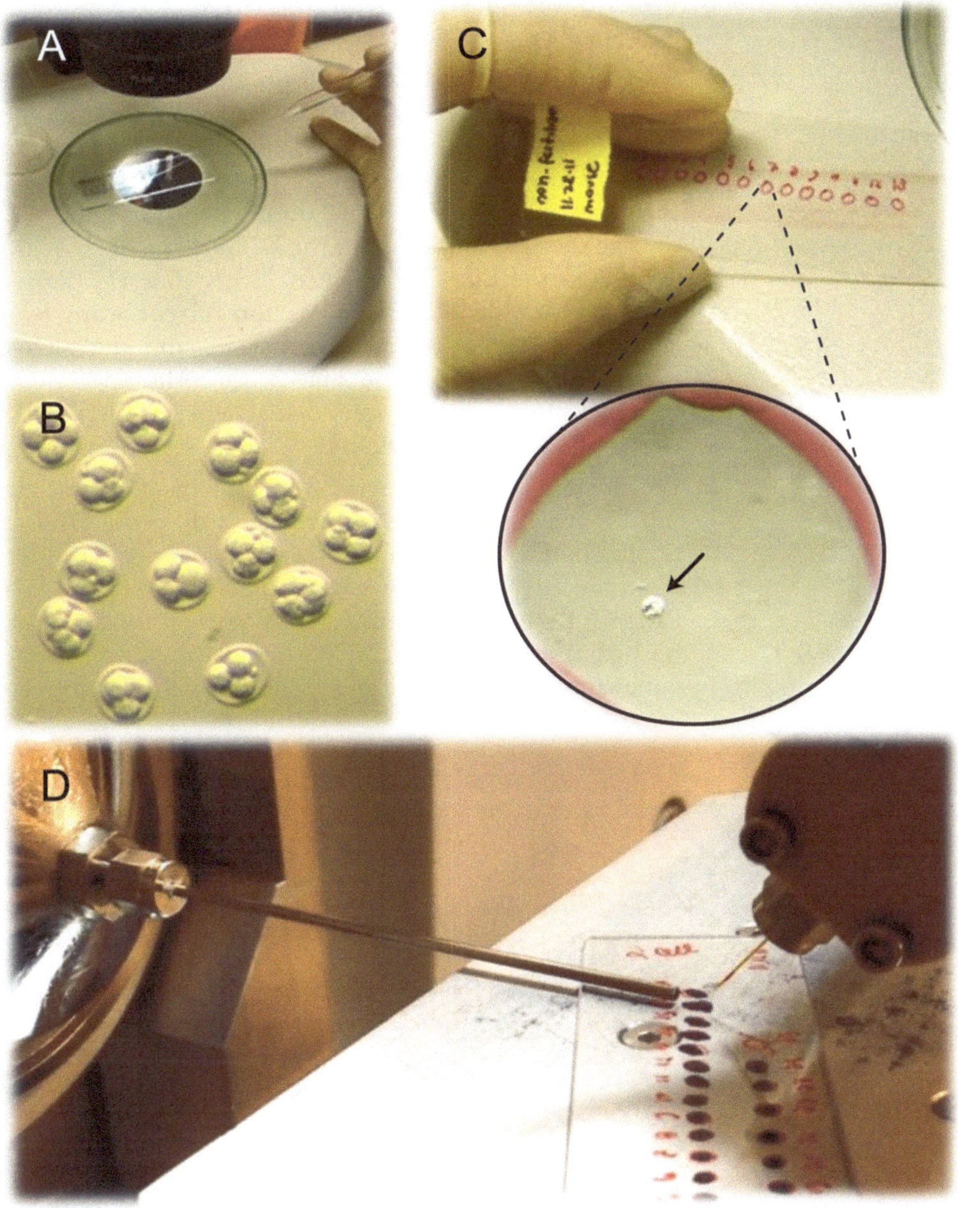

Fig. 1 Overview of sample handling for DESI-MS analysis of individual oocytes/preimplantation mouse embryos. (**a**, **b**) Samples washed under a stereomicroscope in PBS + 0.1% PVA and in methanol/water 1:1 (v/v) and (**c**) individually placed on glass slides. Arrow shows one embryo (circa 100 μm diameter) inside a red ink circle drawn in the back of the slide. (**d**) Lipid profiles recorded by DESI-MS by placing oocytes and embryos under the DESI spray spot. Optical imaging during the experiment shows the DESI spray tip directed toward the glass surface as well as the metal capillary used to conduct the ions to the mass spectrometer. Reproduced with permission from [17]

3. Position the glass slide on the DESI stage optimized as described above in Sect. 3.3.
4. Start MS analysis and data acquisition. First, position the glass slide so that the DESI spray strikes a clean area of glass microscope slide to observe the background and record the MS signal. Then, move the DESI stage or the glass slide out of the spray so that the ion signal drops. Reposition the glass slide to have the first sample under the DESI spray. Remove the glass slide from the DESI spray again. Every time the glass slide is removed from the DESI spray the ion current drops, providing an evident interval between the background and each sample mass spectra, as shown in Fig. 2. One should annotate the order of sample acquisition so that MS scans corresponding to samples can be recognized.
5. When using nondestructive solvent systems, the same embryos can be analyzed multiple times using different DESI conditions, so that a greater number of lipids can be detected. Typically after recording negative ion spectra (i.e., FFA, PS, PE, PC, PG) with DMF/ACN 1:1 (v/v/), the syringe can be washed and refilled with ACN doped with silver nitrate. The same samples can then be analyzed again with the same strategy

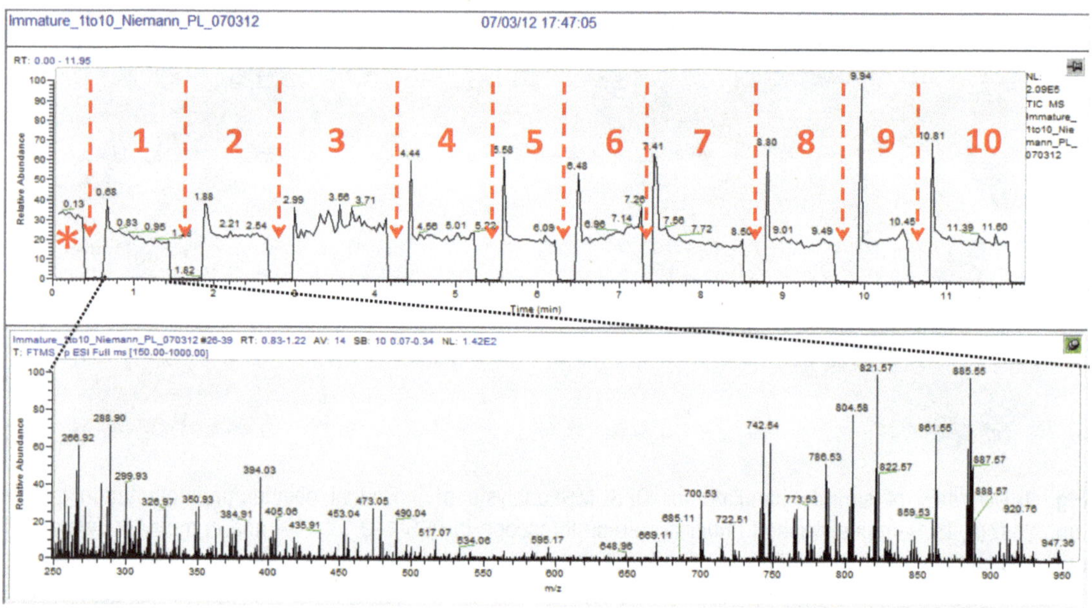

Fig. 2 Chronogram of data acquired from bovine oocytes using an Orbitrap mass spectrometer. (**a**) Chronogram showing the data for ten samples (numbers 1–10). For the data acquisition, first the background of the glass slide is recorded (asterisk). Then the glass slide is removed and no ion signal is observed (dashed arrows). The numbers 1–10 show the ion current during data acquisition for each sample (~1 min/sample). (**b**) Representative negative ion mode high-resolution mass spectrum of an immature oocyte sample. Reproduced from [14] in compliance with Creative Commons Attribution (CC BY) license

discussed above, in order to record lipids such as TAG, DAG, ubiquinone, cholesterol and cholesterol derivatives (cholesterol esters), in positive ionization mode as silver adducts.

6. For some samples, DESI-MS profile has been observed to change within the first couple of minutes due to differential extraction of some lipids [17]. For most samples of mouse oocytes and preimplantation embryos, FFA dimers were more abundantly detected in the first minute while the ion signal for PL predominated later (Fig. 3).

3.5 Data Analysis (See Note 20)

3.5.1 Multivariate and Univariate Statistics

1. Once MS data acquisition is complete, mass spectra can be visualized using the instrument software. Typical DESI-MS mass spectra of mouse oocytes and preimplantation embryos are illustrated in Fig. 4.
2. For each sample, the mass spectra can be averaged by selecting across multiple scans acquired over time (with or without subtraction of the background), and the data exported in Excel as a list of *m/z* values and ion abundances. Full scan mass spectra acquired using a linear ion trap MS (i.e., low mass resolution) typically have a constant step in *m/z*, so each spectrum will consist of the same number of *m/z* values over the same mass range. A unique excel sheet compiling the averaged mass spectra for all the samples can be saved, so that the number of rows corresponds to the number of samples and

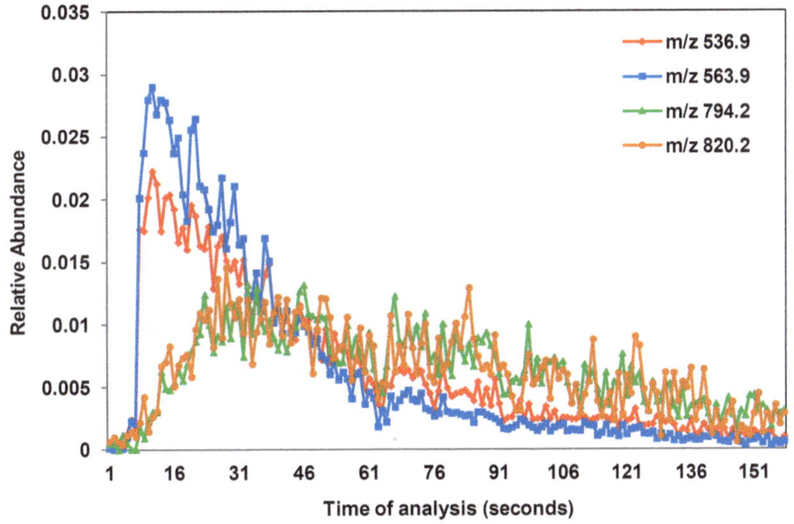

Fig. 3 Relative ion abundance vs. analysis time for two fatty acid dimers (*m/z* 536.9 and *m/z* 563.9) and two phospholipid species (*m/z* 794.2 and *m/z* 820.2) present in a single unfertilized mouse oocyte. In the first minute of analysis, fatty acid dimers are rapidly extracted from the sample by the ACN/DMF spray. Later, ion signals for PL species predominate. (Reproduced with permission from [17])

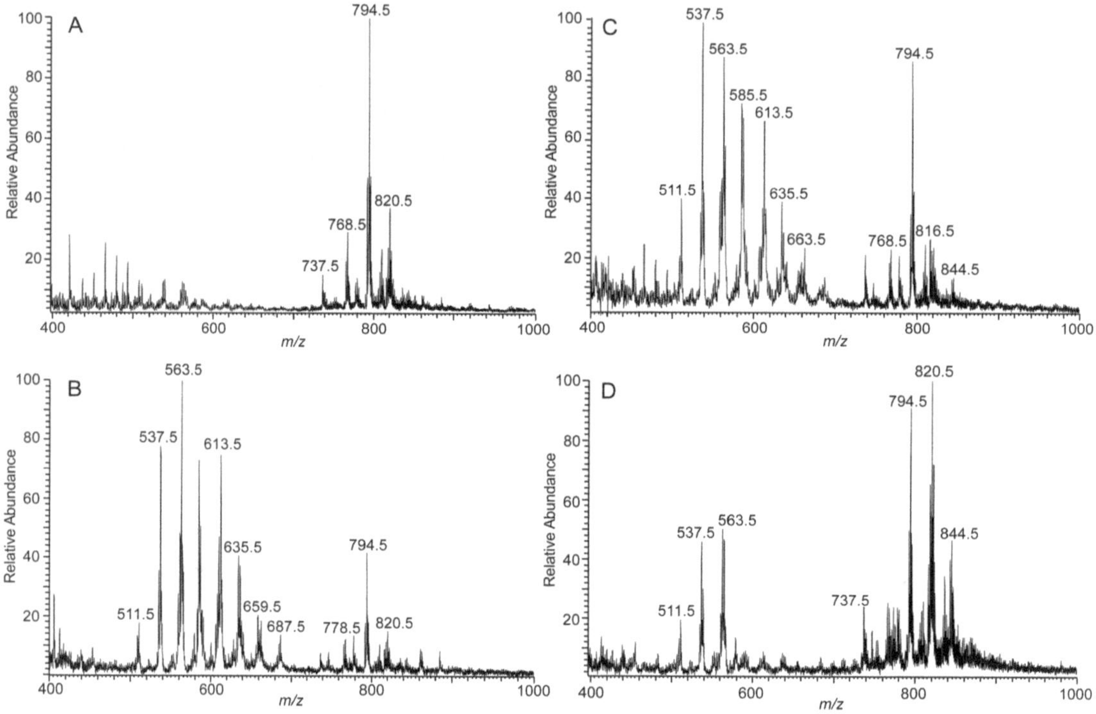

Fig. 4 Representative negative ion mode DESI mass spectra (accumulated over 2–3 min of analysis for each sample) of individual oocytes and embryos from each of four mouse preimplantation developmental stages (**a**) unfertilized oocyte; (**b**) 2-cell embryo; (**c**) 4-cell embryo; (**d**) blastocyst. Reproduced with permission from [17]

the number of columns corresponds to the *m/z* values. This is the basic structure required to upload the MS data into commercial software for statistical analysis such as MATLAB® (The Math Works, Inc., Natick, USA). For high-resolution mass spectra, resampling is usually necessary as a preprocessing step in order to compare spectra from different samples, as the number of *m/z* values and the number between consecutive *m/z* values vary [18].

3. If interested in examining the abundance of a single lipid, the maximum intensity of the *m/z* value (or the peak area) corresponding to the analyte of interest should be compared using univariate statistics. If performing MS^n the product ions intensity can be compared. Otherwise, the entire mass spectrum acquired in full scan mode can be considered as a chemical fingerprint of the sample (e.g., lipid profile) and multivariate statistics can be used for pattern recognition between samples and variables (*m/z* values). Many techniques based in multivariate statistics are available for pattern recognition and exploratory analysis (*see* **Note 21**).

3.6 Lipid Attribution

3.6.1 Low Mass Resolution Data (See Note 22 and Fig. 5)

1. After identifying a mass spectral peak of interest, collect MS/MS data on the precursor ion by isolating it using the scan mode settings in the mass spectrometer software. A peak can be deemed interesting simply because the compound is expected or targeted or based on observed intensity differences from different sample groups (e.g., oocytes grown in different media or under altered/hypoxic conditions), a statistical result as described above. The precursor ion should be isolated to as many significant figures available from the data or simulations (e.g., *m/z* 303 vs. *m/z* 303.23 for the deprotonated ion ($[M\text{-}H]^-$) of arachidonic acid detected in negative mode) with 0 collision energy (arbitrary units).
2. After observing successful precursor isolation, increase the collision energy in a stepwise fashion from low (~15 arbitrary units, *see* **Note 23**) to moderate (20–25 arbitrary units) to high (30–40 arbitrary units) to observe product ions. The

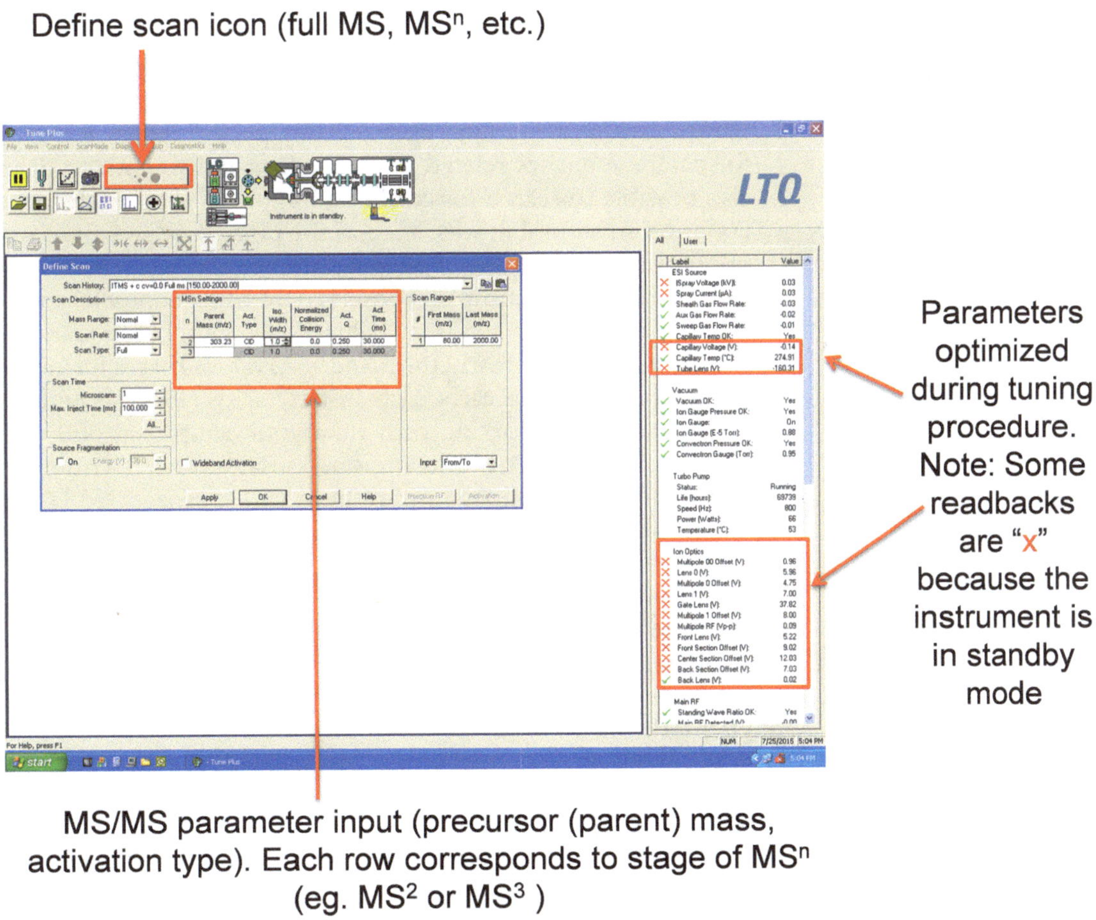

Fig. 5 Photograph of the LTQ Tune GUI showing fields to specify scan parameters needed for MS^n experiments and tuning parameters

amount of collision energy required to observe ion fragmentation is related to the stability of the ion generated and bonds. It is recommended to maintain the instrument's default isolation Mathieu q value and activation time for CID, but these values can be optimized. A good MS/MS spectrum should retain ~5–10% relative abundance of the precursor ion while yielding product ions with a stable intensity and consistent ratio between product ions. Product ions of greater intensity are indicative of more thermodynamically stable structures that result from breaking weaker bonds in the precursor. If using an ion trap mass spectrometer, abundant products from the MS^2 experiment can be isolated and fragmented in an additional dissociation experiment. If conducting the DESI experiment with $AgNO_3$ in the spray solvent as described above, then the $[M+^{109}Ag]^+$ precursor (observed at 2 mass units higher than monoisotopic precursor, $[M+^{107}Ag]^+$, and approximately equal abundance) can be isolated and fragmented to provide confirmatory information. The only difference in the MS^2 spectrum for the heavy silver isotope adduct (i.e. 109 Ag versus 107 Ag) should be an increase of the product ions by 2 mass units.

3. Identification of the lipid class is easily performed by looking for database entries related to *m/z* values and fragments. It is also possible to search for common fragments that result from common structural motifs, such as the polar headgroups (e.g., phosphocholine or phosphoethanolamine). Compute the neutral loss by subtracting the observed product ion *m/z* value from the precursor ion *m/z* value. The neutral loss can be compared to a list of structures and *m/z* values that is included in Fig. 6. Only even carbon chain fatty acids (FAs) should be considered when working with eukaryotic samples as odd carbon chain lengths are uncommon. Unsaturation is common and can be determined using the same approach. MS/MS spectra of analytical standards can be used for confirmation as well as database searches, such as LipidMaps, METLIN, and the Human Metabolome Database (HMBD) (*see* **Note 24**). The interested reader is directed to literature regarding structural assignment of lipids based on nano-electrospray ionization due to similarities in ionization mechanism to DESI [19, 20].

3.6.2 High Mass Resolution Data (See **Note 25**)

1. The measured mass should be input in the METLIN database (https://metlin.scripps.edu), which is freely available at the provided link. The mass error (in ppm) of matching lipids should be less than 5 ppm and can routinely be lowered with a good calibration. The applied ppm mass error filters, and charge state of the ion should be selected from the menu. For

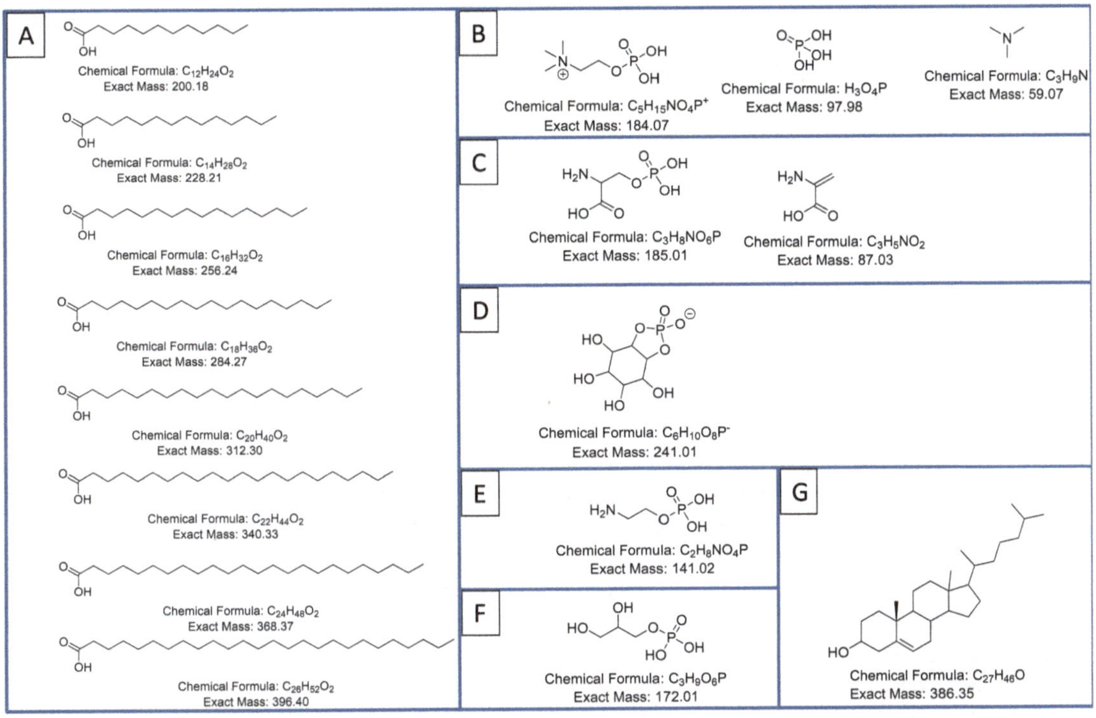

Fig. 6 Structures of fatty acids, ions and characteristic neutral losses observed in CID of various lipid species. (**a**) FFAs from all lipid species discussed, (**b**) PCs, (**c**) PSs, (**d**) PIs, (**e**) PEs, (**f**) PGs, and (**g**) cholesterol derivatives [19, 20]

operation in the negative mode where the ion is presumed to be detected as the deprotonated molecule, the exact mass of the proton (1.00728) can be added to the m/z value and the mass searched for under the neutral filter. The stringency applied by the ppm mass error tolerance is dictated by the performance of the instrument used to collect data. An additional free online resource for searching lipid exact masses or candidate formulas is LIPID MAPS (http://www.lipidmaps.org/). The candidate list of molecular formulas can be corroborated with the neutral losses observed for that ion from the experiments on the ion trap. The combination of molecular formula, compound class (e.g., PC, PI, CE), and fatty acid losses can be used to propose tentative lipid attributions with high confidence.

2. Standards of the most probable structures as determined from molecular formula and fragmentation profiles should be purchased and analyzed under identical MS conditions to validate the attribution (Fig. 7). This may be costly if the expected lipid is particularly exotic but necessary if unambiguous identification of the species of interest is required.

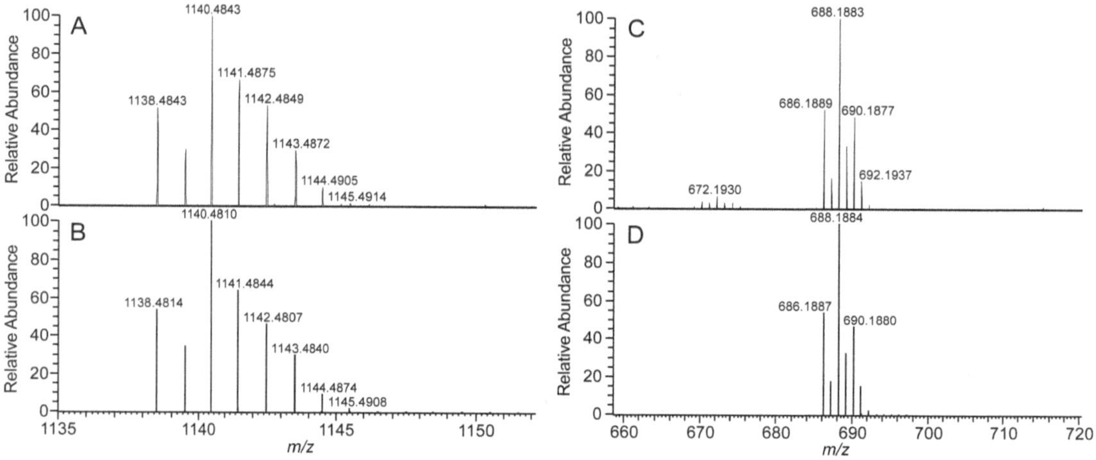

Fig. 7 Attribution of ubiquinone and squalene based on high-resolution mass attribution and in the use of analytical standards to predict formula and isotopic pattern using the Xcalibur software (Thermo Scientific, San Jose, CA, US). These lipids can be detected as silver adducts by DESI-MS and present complex isotopic profiles. (**a**) High mass resolution DESI-MS mass spectrum of the ubiquinone standard showing the region of *m/z* 1135-1155, and (**b**) the correspondent theoretical isotopic distribution of the predicted formula $C_{59}H_{90}O_7NAg2$ or ubiquinone + Ag_2NO_3 (ppm error: 2.568). (**c**) High-resolution DESI-MS mass spectrum of the squalene standard showing the region of *m/z* 660–720, and (**d**) the correspondent theoretical isotopic distribution of $C_{30}H_{50}O_3NAg_2$ or squalene + Ag_2NO_3 (ppm error: 0.153)

3.7 Comments on DESI-MS Analysis Particularities and the Biological Interpretation of Lipid Profiles

DESI-MS provides lipid profiling of individual microscopic oocytes and preimplantation embryos in its native state (in situ analysis). Other MS-based analytical approaches for the analysis of intact lipids such as phospholipids (PL) and triacylglycerols (TAG) including sample extraction and separation may require pooling of tens to hundreds of oocytes or embryos [15, 21, 22].

An overview of the outcomes of DESI-MS lipid analyses and the biological insights derived from that has been presented recently in Ferreira et al. [21]. Differences in the lipid content were found between diverse (1) species or breeds, (2) oocyte maturation and embryo developmental stages, (3) cryopreservation conditions, (4) environmental growth conditions, viz., supplementing the long-term diet with specific nutrients.

For the biological interpretation of differences observed in the lipid profile, it is important first to consider the main biological function of the lipid class. For example, PC determines the cellular membrane bilayer structure while the main role of TAG is cellular energy storage and PIs are mainly signaling lipids. The unsaturation and fatty acyl chain sizes of different lipids can be correlated with the lipid class function. As an illustrative example, the presence of unsaturated and short-chain PC can be correlated with higher membrane fluidity, higher metabolic rates, and higher membrane resistance to cryopreservation. Lipids present in the DESI-MS spectrum can be further correlated by pathway analysis using metabolomics software.

It is critical to keep in mind that DESI-MS lipid profiles correspond to end-points in metabolism (or the metabolic condition at the time of sampling). Changes in lipid profiles of structural lipids depend on metabolic upstream events (gene expression and protein activity) as well as on lipid degradation, turnover, and de novo synthesis. Therefore, enough time between delivering the stimulus to the oocyte or to the embryo (such as supplementing the culture media of oocytes) and sample collection for MS analysis should be allowed to observe changes in the lipid profiles. On the other hand, changes in signaling lipids, such as PI dynamics after fertilization, are expected to occur in short windows (minutes) and sampling at the right time becomes challenging since fertilization occurs in a larger time window after oocytes are exposed to sperm cells. Pilot experiments are always helpful to determine the best timing of the experiment.

The fact that the DESI-MS is an end-point metabolic analysis should be noted when correlating DESI-MS data to upstream metabolism such as gene or protein expression. This means that gene and protein changes are expected some hours or even days after (depending on the pathway, metabolic rate, and stimulus intensity) significant changes in the lipid profiles occur. The changes that can be observed in the lipid profiles depend on the lipid class. Membrane structural lipids are not expected to change greatly in a short period since membranes are well-conserved structures that must remain stable for proper cell function and survival. Nonetheless, we have observed radical changes in membrane composition at early development, probably related to cell differentiation and specialization [14, 17]. Profiles of storage lipids (TAG) and ubiquinone (indicators of mitochondrial activity) can change radically during early development [14].

4 Notes

1. Extra samples (test samples) are necessary for signal and DESI spray optimization. Other lipid-rich samples, such as brain tissue, can be used for this purpose instead. Depending on the instrument settings and sensitivity, the signal obtained from one single oocyte or embryo may not be detectable. The optimization should be checked each time if the spray angle or some other setting is altered. This occurs constantly in multiuser instruments. The condition of the instrument (contamination, background signal) may interfere with sensitivity, so that test samples will guarantee that good quality data will be obtained from the experimental samples. For multiuser instruments, data from all samples of one experiment or at least

part of the samples from every group should be taken in the same day to avoid day and instrument condition-associated variations.

2. Depending on the species, different amounts of lipids occur. As examples, pig oocytes are very rich in TAG, while mouse oocytes are not. Also, extensive variation in the lipid profiles has been observed due to developmental stage and culture conditions [21].
3. Lysis has been observed in some samples stored in microtubes that were frozen (unpublished data). We recommend using phosphate buffer saline (PBS) supplemented with 1% polyvinyl alcohol (PVA) and siliconized microtubes for storage. Frozen samples are to be stored only for short periods of time (~1 month).
4. These are mouth pipettes or precision pipettes commonly used in embryology labs. There is no preference for a specific type, but the pipette should allow transferring the embryo from the culture plate to droplets where it will be washed and subsequently transferred to the glass slide for DESI-MS analysis.
5. Unpublished data indicate that siliconized tubes are important to avoid sample lysis. Volume of 0.6 mL is suggested but microtubes of other volumes can be used.
6. These are plain glass slides, not coated. Cleaned glass slides should be used to avoid background signal or the presence of contaminating molecules that can disturb the lipid profiles, such as polymers. One protocol to clean glass slides is sonicate them for at least 10 min in pure acetone.
7. Fine-point Sharpie® markers are recommended.
8. Silver nitrate should be of high purity such as silver nitrate 99.9999% (Sigma ref. #204390). Degradation (reduction to solid silver and formation of silver nanoparticles) of this reagent is likely to occur with time due to air and light exposure. This can cause increased background noise in the mass spectra and a decrease in the cationization efficiency.
9. DESI sources can be home-built using a stainless steel union tee and silica capillaries (Fig. 8). Commercial DESI sources are available from Prosolia Inc. To our knowledge these commercial sources have never been used for the analysis of oocytes and preimplantation embryos, but should be just as suitable.
10. To the best of our knowledge, so far linear ion traps and Orbitrap mass analyzers have been used for the lipid analysis of single oocytes and embryos. Low resolution mass spectra can be informative if structural information can be obtained by MS/MS. More details are explained in the topic "lipid attribution" in this chapter.

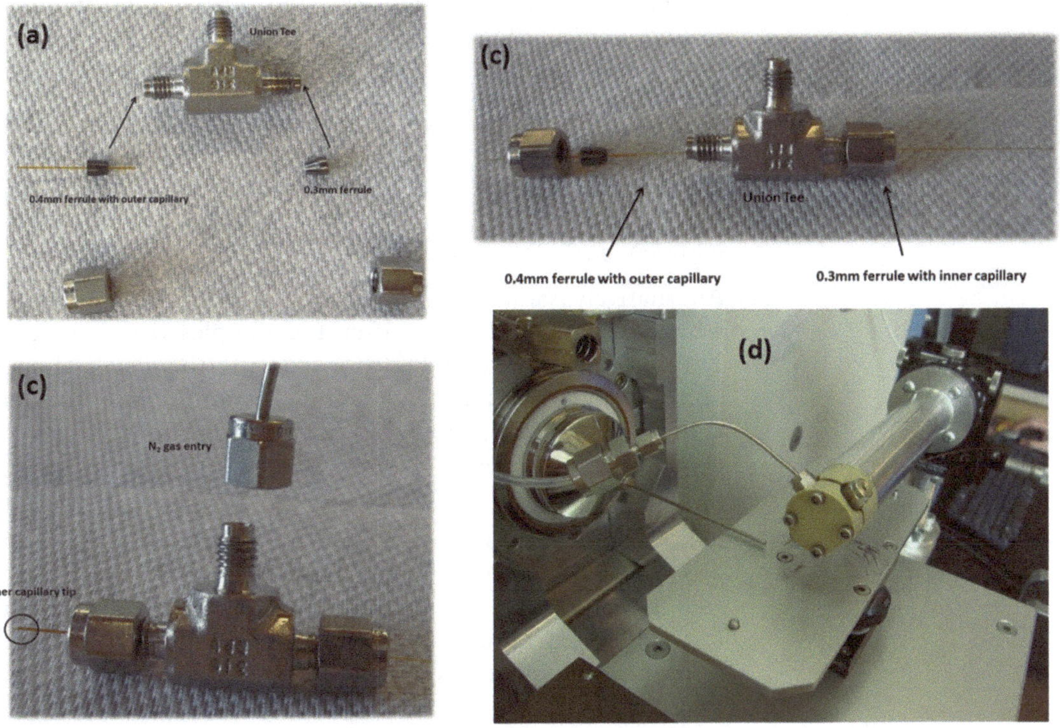

Fig. 8 Tutorial on building a DESI spray. Needed components are (Swagelok): Tube Fitting, Union Tee, 1/16 in. Ferrules (Restek): Restek P/N 20233 1/16″ × 0.3 mm graphite (fix the inner capillary in the metal piece), and Restek P/N 20229 1/16″ × 0.4 mm vespel/graphite (fix the outer capillary in the metal piece). Capillaries (Polymicro): Outer capillary TSP250350 and Inner capillary TSP050150. (**a**) Initial assembly; (**b**) Fixing inner and outer capillary; (**c**) N_2 entry adapter; (**d**) Home-built spray to optimization. More information about DESI spray optimization and setting can be found on Eberlin 2014 [20]

11. A simple electric desiccator is employed, such as a Cole-Parmer Desi-Vac model 3164. We do not recommend drying the sample too much. Simply leaving the samples to dry on the glass slide at the air-conditioned ambience of a laboratory for circa 30 min should be suitable if no desiccator is available.
12. Most mass spectrometers have their own precision pump and compatible glass syringes for solvent delivery. Syringe with volumes of 250 or 500 μL are appropriated for the flow rates used for the negative ion mode (solvent 1:1 DMF:ACN, usually 1 μL/min) and for the positive ion mode (5 ppm $AgNO_3$ in ACN, 4–6 μL/min).
13. The washing solutions are stable and can be prepared and stored in advance.
14. Marking sample position with a black fine-point marker (and with a number for sample identification) on the back of the glass slide is helpful since they are tiny samples. The marking can be done before or after the placement of the samples onto

the glass slide. We prefer doing it before so that one can use the marks to guide where samples are placed.

15. This step is intended to remove culture medium lipids.
16. Due to fast methanol evaporation during sample wash, oocytes and embryos may move inside the droplet. Perform washing in this solution for just a few seconds. The purpose of this washing step is to remove culture media/PBS salts that will interfere with MS analysis.
17. Alternatively, samples can be placed onto the glass slides after being exposed to SWS1. After the samples dry completely, 1 mL of SWS2 is carefully deposited onto the glass slides and removed by orienting the glass slide vertically. This procedure is more risky since some samples can be washed away or can change their positions.
18. Other histologically compatible solvents such as DMF combinations with ethanol, chloroform, and acetonitrile [10] would probably be suitable for oocyte and embryo lipid profiling as well. In some samples the zona pellucida represented a barrier to the lipid extraction by DESI-MS. For such cases, the addition of 1% of formic acid to the ACN/DMF 1:1 (v/v) solvent improved data quality. But mass spectra showed formate adducts of lipids, such as PC (34:1) + HCOO−, of *m/z* 804.5 [16].
19. The concentration of silver nitrate in acetonitrile may be optimized. This solvent combination should be prepared shortly before the experiment. We have observed additional background signals if acetonitrile supplemented with 3.5–5 ppm silver nitrate is kept at room temperature for more than 1 day.
20. Data handling and statistical approaches based on recognitions of patterns (i.e., lipid profiles) can be used to characterize differences in chemical content between oocytes and embryos. We present a general workflow and for details please refer to Ferreira et al. [21]. General overview on pattern recognition strategies are described elsewhere [23].
21. Using multivariate statistics it is possible to extract spectral information from the full MS data matrix and account for the inter-correlations among ion abundances and *m/z* values. Multivariate statistics techniques can be classified generally as unsupervised or supervised methods. The first provides information about the relationships among the samples and the variables studied, whereas the latter adds mathematical models for the prediction of qualitative or quantitative features when unknowns are present. Additional data preprocessing and variable selection techniques are commonly combined to multivariate statistics [24].

Table 2
Examples of typical fragments and neutral losses from animal lipids [19], which are commonly observed in oocytes and embryos

Precursor or neutral loss fragment	Lipid class
m/z 184 (+) and NL 59 (+)	PC/LysoPC/SM
NL 141	PE/LysoPE
NL 185	PS
NL 277	PI
NL 189	PG
NL 115	PA
Prec 153 (−)	LysoPG

22. Considered alone, full-scan MS data generated using a low mass resolving power instrument ($m/\Delta m = 1000$) such as a linear ion trap is insufficient to assign a tentative structure to an ion. But the use of tandem mass spectrometry (MS/MS or MS^2) can provide broad classification of lipids based on characteristic neutral losses (*see* Table 2) observed in collisional induced dissociation (CID). Other fragmentation techniques can provide different characteristic losses and may allow for more precise structural elucidation (e.g., stereochemistry of double bonds using ozone, position on glycerol backbone using lithium adduction [25]). Structural characterization of lipids observed in oocytes and preimplantation embryos has been reviewed in detail recently [21].
23. These values of collision energy have been observed for in a LTQ Thermo Scientific using helium as the collision gas (San Jose, CA).
24. LipidMaps: http://www.lipidmaps.org/ and METLIN: https://metlin.scripps.edu/index.php
25. With proper calibration, high-resolution (mass resolving power over 100,000) data generated from an Orbitrap mass spectrometer can be used to determine a candidate list of molecular formulas derived from a measured *m/z* value. It is imperative to operate with the exact monoisotopic mass of elements when working with strict ppm mass error limits. The free program IsoPro 3.1 (https://sites.google.com/site/isoproms/home/download) is recommended for calculating monoisotopic masses and simulating the isotopic distributions of candidate formulas. The experimental isotopic distribution should match the simulated one if the candidate molecular formula corresponds to the actual ion, so isotope matching is an additional confirmatory procedure.

Acknowledgments

This work was supported by NSF CHE 1307264. C.R.F. was supported from the Purdue University Center for Cancer Research Small Grants and from the Brazilian National Council for Scientific and Technological Development (CNPq). The funders had no role in study design, data collection and analysis, decision to publish, or preparation of the manuscript.

References

1. Cooks RG, Ouyang Z, Takats Z, Wiseman JM (2006) Detection technologies. Ambient mass spectrometry. Science 311(5767):1566–1570. https://doi.org/10.1126/science.1119426. 311/5767/1566 [pii]
2. Badu-Tawiah AK, Eberlin LS, Ouyang Z, Cooks RG (2013) Chemical aspects of the extractive methods of ambient ionization mass spectrometry. Annu Rev Phys Chem 64:481–505. https://doi.org/10.1146/annurev-physchem-040412-110026
3. Takats Z, Wiseman JM, Gologan B, Cooks RG (2004) Mass spectrometry sampling under ambient conditions with desorption electrospray ionization. Science 306 (5695):471–473. https://doi.org/10.1126/science.1104404. 306/5695/471 [pii]
4. Monge ME, Harris GA, Dwivedi P, Fernandez FM (2013) Mass spectrometry: recent advances in direct open air surface sampling/ionization. Chem Rev 113(4):2269–2308. https://doi.org/10.1021/cr300309q
5. Cooks RG, Jarmusch AK, Ferreira CR, Pirro V (2015) Skin molecule maps using mass spectrometry. Proc Natl Acad Sci U S A 112 (17):5261–5262. https://doi.org/10.1073/pnas.1505313112. 1505313112 [pii]
6. Santagata S, Eberlin LS, Norton I, Calligaris D, Feldman DR, Ide JL, Liu X, Wiley JS, Vestal ML, Ramkissoon SH, Orringer DA, Gill KK, Dunn IF, Dias-Santagata D, Ligon KL, Jolesz FA, Golby AJ, Cooks RG, Agar NY (2014) Intraoperative mass spectrometry mapping of an onco-metabolite to guide brain tumor surgery. Proc Natl Acad Sci U S A 111 (30):11121–11126. https://doi.org/10.1073/pnas.1404724111. 1404724111 [pii]
7. Siebenhaar M, Kullmer K, de Barros Fernandes NM, Hullen V, Hopf C (2015) Personalized monitoring of therapeutic salicylic acid in dried blood spots using a three-layer setup and desorption electrospray ionization mass spectrometry. Anal Bioanal Chem. https://doi.org/10.1007/s00216-015-8887-8
8. Eberlin LS, Gabay M, Fan AC, Gouw AM, Tibshirani RJ, Felsher DW, Zare RN (2014) Alteration of the lipid profile in lymphomas induced by MYC overexpression. Proc Natl Acad Sci U S A 111(29):10450–10455. https://doi.org/10.1073/pnas.1409778111. 1409778111 [pii]
9. Talaty N, Mulligan CC, Justes DR, Jackson AU, Noll RJ, Cooks RG (2008) Fabric analysis by ambient mass spectrometry for explosives and drugs. Analyst 133(11):1532–1540. https://doi.org/10.1039/b807934j
10. Eberlin LS, Ferreira CR, Dill AL, Ifa DR, Cheng L, Cooks RG (2011) Nondestructive, histologically compatible tissue imaging by desorption electrospray ionization mass spectrometry. Chembiochem 12(14):2129–2132. https://doi.org/10.1002/cbic.201100411
11. Campbell DI, Ferreira CR, Eberlin LS, Cooks RG (2012) Improved spatial resolution in the imaging of biological tissue using desorption electrospray ionization. Anal Bioanal Chem 404(2):389–398. https://doi.org/10.1007/s00216-012-6173-6
12. Cooks RG, Manicke NE, Dill AL, Ifa DR, Eberlin LS, Costa AB, Wang H, Huang G, Ouyang Z (2011) New ionization methods and miniature mass spectrometers for biomedicine: DESI imaging for cancer diagnostics and paper spray ionization for therapeutic drug monitoring. Faraday Discuss 149:247–267. discussion 333–256. https://doi.org/10.1039/C005327A
13. Eberlin LS, Ferreira CR, Dill AL, Ifa DR, Cooks RG (2011) Desorption electrospray ionization mass spectrometry for lipid characterization and biological tissue imaging. Biochim Biophys Acta 1811(11):946–960. https://doi.org/10.1016/j.bbalip.2011.05.006. S1388-1981(11)00073-4 [pii]
14. Gonzalez-Serrano AF, Pirro V, Ferreira CR, Oliveri P, Eberlin LS, Heinzmann J, Lucas-Hahn A, Niemann H, Cooks RG (2013) Desorption electrospray ionization mass

spectrometry reveals lipid metabolism of individual oocytes and embryos. PLoS One 8(9): e74981. https://doi.org/10.1371/journal.pone.0074981. PONE-D-13-22791 [pii]

15. Ferreira CR, Eberlin LS, Hallett JE, Cooks RG (2012) Single oocyte and single embryo lipid analysis by desorption electrospray ionization mass spectrometry. J Mass Spectrom 47 (1):29–33. https://doi.org/10.1002/jms.2022
16. Eberlin LS (2014) DESI-MS imaging of lipids and metabolites from biological samples. Methods Mol Biol 1198:299–311. https://doi.org/10.1007/978-1-4939-1258-2_20
17. Ferreira CR, Pirro V, Eberlin LS, Hallett JE, Cooks RG (2012) Developmental phases of individual mouse preimplantation embryos characterized by lipid signatures using desorption electrospray ionization mass spectrometry. Anal Bioanal Chem 404(10):2915–2926. https://doi.org/10.1007/s00216-012-6426-4
18. Pirro V, Eberlin LS, Oliveri P, Cooks RG (2012) Interactive hyperspectral approach for exploring and interpreting DESI-MS images of cancerous and normal tissue sections. Analyst 137(10):2374–2380. https://doi.org/10.1039/c2an35122f
19. Brugger B, Erben G, Sandhoff R, Wieland FT, Lehmann WD (1997) Quantitative analysis of biological membrane lipids at the low picomole level by nano-electrospray ionization tandem mass spectrometry. Proc Natl Acad Sci U S A 94(6):2339–2344. https://doi.org/10.1073/pnas.94.6.2339
20. Schneiter R, Brugger B, Sandhoff R, Zellnig G, Leber A, Lampl M, Athenstaedt K, Hrastnik C, Eder S, Daum G, Paltauf F, Wieland FT, Kohlwein SD (1999) Electrospray ionization tandem mass spectrometry (ESI-MS/MS) analysis of the lipid molecular species composition of yeast subcellular membranes reveals acyl chain-based sorting/remodeling of distinct molecular species en route to the plasma membrane. J Cell Biol 146(4):741–754. https://doi.org/10.1083/jcb.146.4.741
21. Ferreira CR, Jarmusch AK, Pirro V, Alfaro CM, Gonzalez-Serrano AF, Niemann H, Wheeler MB, Rabel RA, Hallett JE, Houser R, Kaufman A, Cooks RG (2015) Ambient ionisation mass spectrometry for lipid profiling and structural analysis of mammalian oocytes, preimplantation embryos and stem cells. Reprod Fertil Dev. https://doi.org/10.1071/RD14310. RD14310 [pii]
22. Ferreira CR, Saraiva SA, Catharino RR, Garcia JS, Gozzo FC, Sanvido GB, Santos LF, Lo Turco EG, Pontes JH, Basso AC, Bertolla RP, Sartori R, Guardieiro MM, Perecin F, Meirelles FV, Sangalli JR, Eberlin MN (2010) Single embryo and oocyte lipid fingerprinting by mass spectrometry. J Lipid Res 51 (5):1218–1227. https://doi.org/10.1194/jlr.D001768. jlr.D001768 [pii]
23. Issaq HJ, Van QN, Waybright TJ, Muschik GM, Veenstra TD (2009) Analytical and statistical approaches to metabolomics research. J Sep Sci 32(13):2183–2199. https://doi.org/10.1002/jssc.200900152
24. Oliveri P, Casale M, Casolino MC, Baldo MA, Grifi FN, Forina M (2011) Comparison between classical and innovative class-modelling techniques for the characterisation of a PDO olive oil. Anal Bioanal Chem 399 (6):2105–2113. https://doi.org/10.1007/s00216-010-4377-1
25. Murphy RC, Axelsen PH (2011) Mass spectrometric analysis of long-chain lipids. Mass Spectrom Rev 30(4):579–599. https://doi.org/10.1002/mas.20284
26. Pirro V, Oliveri P, Ferreira CR, Gonzalez-Serrano AF, Machaty Z, Cooks RG (2014) Lipid characterization of individual porcine oocytes by dual mode DESI-MS and data fusion. Anal Chim Acta 848:51–60. https://doi.org/10.1016/j.aca.2014.08.001. S0003-2670(14)00948-9 [pii]
27. Gonzalez-Serrano AF, Ferreira CR, Pirro V, Lucas-Hahn A, Heinzmann J, Hadeler KG, Baulain U, Aldag P, Meyer U, Piechotta M, Jahreis G, Danicke S, Cooks RG, Niemann H (2015) Effects of long-term dietary supplementation with conjugated linoleic acid on bovine oocyte lipid profile. Reprod Fertil Dev. https://doi.org/10.1071/RD14352

Chapter 14

Spatial Mapping of Cellular Metabolites Using DESI Ion Mobility Mass Spectrometry

Anthony Midey, Hernando Olivos, and Bindesh Shrestha

Abstract

Spatial mapping of cellular metabolites, such as neurotransmitters and lipids, on the tissue, can increase our understanding of the biological functions of those molecules. Mass spectrometry imaging (MSI) techniques, such as desorption electrospray ionization (DESI), have not demonstrated the ability to perform metabolite analysis at mammalian single cell level yet. However, they can be a valuable tool to provide insight into cellular metabolism in a very small population (tens) of cells. DESI MSI, coupled with ion mobility separation, improves the peak capacity and signal-to-noise ratio of detected analytes by separating a molecule of interest from interfering isobaric species found in a complex biological matrix. Here we present a protocol for mapping cellular metabolites neurotransmitters, such as serotonin, adenosine, and glutamine directly in brain tissue samples using DESI MSI.

Key words Metabolite imaging, DESI, Ion mobility, Collision cross section, Traveling-wave ion mobility

1 Introduction

Mass spectrometry imaging (MSI) can be used to map the distribution of molecules in tissue comprised of small populations of cells. Desorption Electrospray Ionization (DESI) is an ambient ionization technique that often provides complementary molecular information to other MSI techniques, such as matrix-assisted laser desorption ionization (MALDI). One of the significant challenges in metabolic imaging is preferential ionization of lipids species. Tissue can be washed to reduce the intensity of lipid species, but the same workflow will also wash off many metabolites. Thus, the alternative approach will be to either use derivitization of metabolite species to enhance their analysis or use complementary imaging techniques more amenable to metabolites detection. DESI MSI has been employed to detect cellular metabolites from mammalian tissue sections, microbial colonies, cell cultures, etc. [1, 2].

Bindesh Shrestha (ed.), *Single Cell Metabolism: Methods and Protocols*, Methods in Molecular Biology, vol. 2064,
https://doi.org/10.1007/978-1-4939-9831-9_14,

Detection and annotation of metabolites are enhanced by using ion mobility separation (IMS) in concert with DESI imaging. Ion mobility is a gas-phase electrophoretic separation technique that can enhance the detection of low abundant molecular species by separating them from highly abundant background ions. Ion mobility separates ions in the gas phase after ionization according to their charge, shape, and size in milliseconds, making it compatible with the fast pixel-by-pixel acquisition of any mass spectrometry imaging techniques. In addition to enhancing S/N ratio, drift times of ions can be converted into collision cross section (CCS) and matched with a database for improving metabolite annotation. DESI-IMS-MS imaging experiments measured a mass spectrum (intensity vs. mass-to-charge, *m/z* vs., drift time) at each pixel on a virtual grid overlaying analyzed.

Four IMS-MS technologies are currently commercially available: drift-time ion mobility spectrometry (DT-IMS), trapped ion mobility, differential-mobility spectrometry (DMS) or field–asymmetric ion mobility spectrometry (FAIMS), traveling-wave ion mobility. DESI-MS imaging with ion mobility is supported only in one commercial product, such as SYNAPT G2-XS mass spectrometer traveling-wave ion mobility cell or high definition mass spectrometry (HDMS).

2 Materials

2.1 Equipment

1. Desorption Electrospray Source: DESI source (Waters Corporation).
2. Mass spectrometer: a traveling-wave ion mobility quadrupole ToF mass spectrometer (SYNAPT G2-XS, Waters Corporation, Milford, MA).
3. Cryostat (e.g., Leica CM1850).

2.2 Reagents

1. Major Mix IMS/Tof Calibration kit containing Poly-DL-alanine and other small molecules (Waters Corporation, Milford, MA; cat # 186008113).
2. Solvents: water (HPLC-grade), methanol; purchased from Sigma (Saint Louis, Missouri, USA). All solvents were HPLC grade or higher.

2.3 Supplies

1. Fresh-frozen tissue section such as rat brain sections (12 μm thick).
2. Standard microscope glass slides (75 × 25 mm, e.g., VWR cat # 490013).

2.4 Software

1. DriftScope (Waters) is used for extracting regions of interest for different molecules based on selective drift time extraction of mass spectra.
2. High Definition Imaging Software version 1.5 (Waters Corporation, Milford, MA).

3 Methods

1. Setting up TWIM-HDMS.
2. Calculating collision cross sections.
3. DESI-HDMS imaging.

3.1 Setting Up DESI HDMS

General routine maintenance and calibration of the mass spectrometer is needed before the experiment, similar to any MS analysis. Traveling wave ion mobility cell is calibrated with compounds of established CCS values, e.g., poly-DL-alanine, before CCS calculation using drift times [3–6]. The calibrations are needed to correlate CCS to drift times, which may shift due to various parameter settings.

1. Perform routine instrument checks, e.g., detector check, and calibration using IntelliStart in MassLynx. The mass spectrometer using sodium formate in both positive and negative mode. Tune the mass spectrometer (SYNAPT G2S*i*, Waters Corporation, Milford, MA) according to the manufacturer's guideline.
2. Mass accuracy of DESI acquisition can be improved by the elimination of systematic mass drift, using either in-line lock mass. Typically, 200–250 pg/μL leu-enk is dissolved in DESI sprayer solvent to be used as lock-mass. Alternatively, a uniformly distributed endogenous ion or a known ESI background ion can also be used as a lock-mass.
3. Set the mass spectrometer in the mobility TOF mode and leave 1 h to equilibrate.
4. Optimize HDMS parameters to make sure that the entire mass range fits in the ion mobility bin window while avoiding aliasing. The main HDMS parameters that need to be tuned to achieve optimal HDMS conditions are: TrapDC bias (set around 35 for metabolites 65 for lipids), transfer wave velocity (set according to Tof pusher frequency), IMS wave height (ramped around 200–1200 m/s), etc.
5. CCS values measured with nitrogen gas were determined using previously published CCS values for singly charged poly-DL-alanine oligomers as the calibrant [5].
6. CCS calibration can be done in IntelliStart by selecting right calibrant similar to doing the mass calibration.

3.2 Tissue Preparation for DESI Imaging

Tissue sectioning step for DESI imaging of cellular metabolites is very similar to the common sectioning for microscopy. However, a critical step for MSI includes avoiding any chemical processing, e.g., formalin-fixed paraffin-embedding (FFPE), and cross-contamination, e.g., polyethylene glycol (PEG) found in optimal cutting temperature (OCT) compound. The following tissue processing steps are recommended for obtaining optimal sections for DESI imaging:

1. Store fresh or flash frozen tissue section at −80 °C before analysis. Please note the orientation of the tissue (*see* **Note 1**).
2. Avoid using OCT embedding. If need be, sparingly use OCT as a glue to hold the tissue block on to the chuck, avoiding any contact with OCT with a blade or other portion of tissue used for MSI.
3. Alternatively, use 2.5% carboxymethylcellulose (CMC) solution dissolved in deionized water as embedding media.
4. Mount the specimen in the desired orientation, e.g., coronal or sagittal for the brain.
5. Remove the tissue block and attach the block to the sample mount of a cryostat (e.g., CM3050 S, Leica Buffalo Grove, IL).
6. Typically, DESI imaging tissue is sectioned between 10 and 20 μm in thickness depending on the morphology of the tissue. Sectioning can be done around −20 °C or based on manufacturer's recommendation for that type of tissue.
7. After sectioning, mount the section of tissue on a slide. Thaw mount by gently warming the back side of the slide with a gloved-hand.
8. Store thaw-mounted tissue sections in airtight, labeled slide holder placed inside a −80 °C freezer until further use (*see* **Note 2**).
9. Remove frozen tissue sections on a slide, and quickly transfer to a vacuum desiccator, and dry for 15 min or until the tissue is visually dried. Avoid condensation on the tissue.

3.2.1 DESI Optimization

DESI ions are produced by desorption and ionization of analytes on a sample surface by pneumatically assisted electrospray directed toward sample surface. A conceptual schematic of a side view of DESI ion source on a three-dimensional translational stage is shown in Fig. 1. The entire DESI assembly and MS inlet capillary is mounted over a scanning two-dimensional translational stage. The spatial resolution of DESI image depends on the spray impact width (tens to a few of hundred microns) and scanning speed of stage (tens of microns per pixel or scan). Correct setup and optimization of DESI source is critical for obtaining a robust and stable

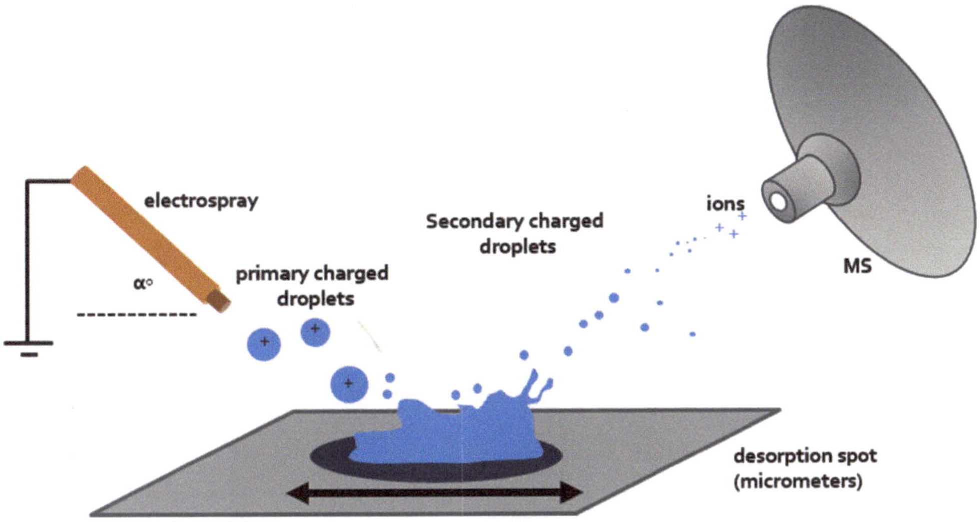

Fig. 1 Conceptual schematic of desorption electrospray ion source

signal. Ideally, optimization is performed with a similar sample, such as adjacent serial sections.

1. Set initial DESI ion source parameters as recommended by manufacturer, such as for nebulizing gas pressures, solvent flow rate, applied voltage, incidence angle (α), the angle of ion collecting inlet capillary (β), the distance between emitter and sample surface ($h_e + h_s$), the distance between the electrospray emitter and MS inlet capillary (d), the distance between ion collection capillary and sample surface (h_s) (*see* **Note 3**).
2. Optimize each set of the sample by monitoring their signal intensity. Initially, optimize MS signal by analyzing a spot drawn by a red Sharpie containing the rhodamine 6G dye (m/z 443) on a glass microscope slide.
3. Begin the optimization process by changing the DESI emitter tip protrusion using a 5-mm wrench. Follow by protrusion alignment by checking spray-direction created on a black surface coated with Staedtler marker. Spray-direction should be toward MS inlet, not away from it. If spray-direction is away, move the entire body of the DESI sprayer to align direction with MS inlet. Repeat (*see* **Note 4**).
4. Electrospray composition for DESI imaging depends on the analyte of interest and tissue. A typical composition of 98% methanol with 0.1% acetic acid (*v/v*) is recommended as a starting point for DESI imaging method development.

3.3 DESI Imaging with Ion Mobility

To obtain DESI images, mass spectra are collected in a series of pixels that are later rendered into false color images based on the intensity of ions. A motorized two-dimension stage with a sample is

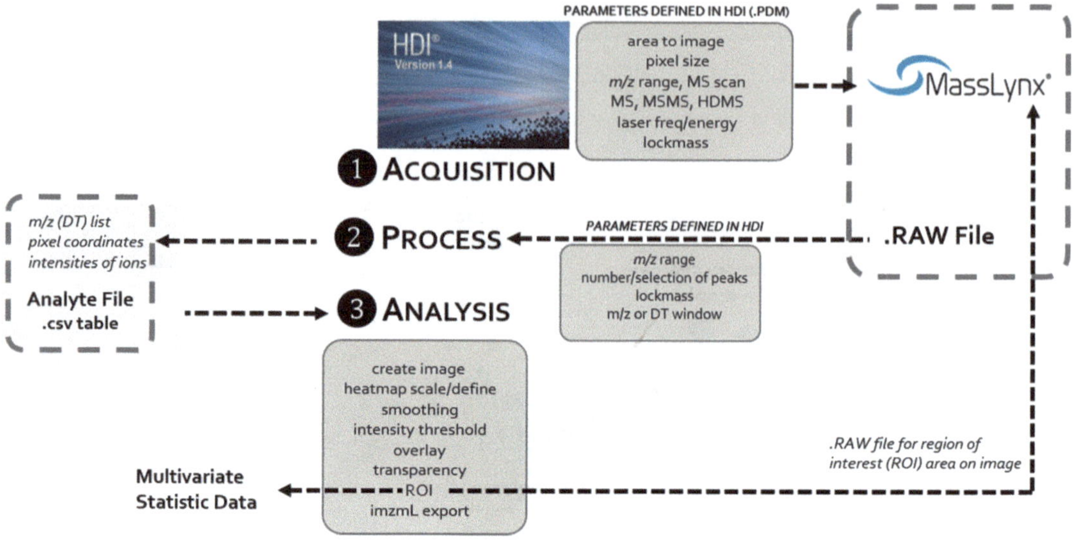

Fig. 2 The workflow of High Definition Imaging Software for acquiring, processing, and analyzing MS i8maging data

moved in subsequent liner lines to obtain mass spectrum at virtually defined pixels.

1. Image co-registration step to define the area for DESI imaging is needed before setting up a DESI experiment. Obtain an optical image of the entire slide using a flatbed scanner or tissue scanner. The four corners of the slide will be used to co-register the slide in High Definition Imaging or HDI (Waters Corporation, Milford, MA) (*see* **Note 5**).
2. In the HDI software, define a rectangular area of DESI-HDMS imaging. The translation stage will move in successive rows creating a two-dimensional image. Define pixel sizes (equal value for x and y is recommended), input desired stage scan speed (e.g., 100 μm/s), and line spacing (e.g., 100 μm) between rows. To obtain data in ion mobility mode ensure that the scan speed of mass spectrometer is not set lower than 0.3 spectra per second—recommended initial setting is 0.5–1 scan/s. A schematic showing the HDI workflow is shown in Fig. 2 (*see* **Note 6**).
3. Set HDI to acquire TWIM-MS imaging data. After the conclusion of the DESI imaging data acquisition, return mass spectrometer to standby mode, turn off nitrogen gas, and syringe pump.
4. The RAW mass spectral data can be processed into ion intensities in each pixel for thousands of ions with the highest overall intensities or defined target list of ions. This processing can be done right with the acquisition or afterward in HDI.

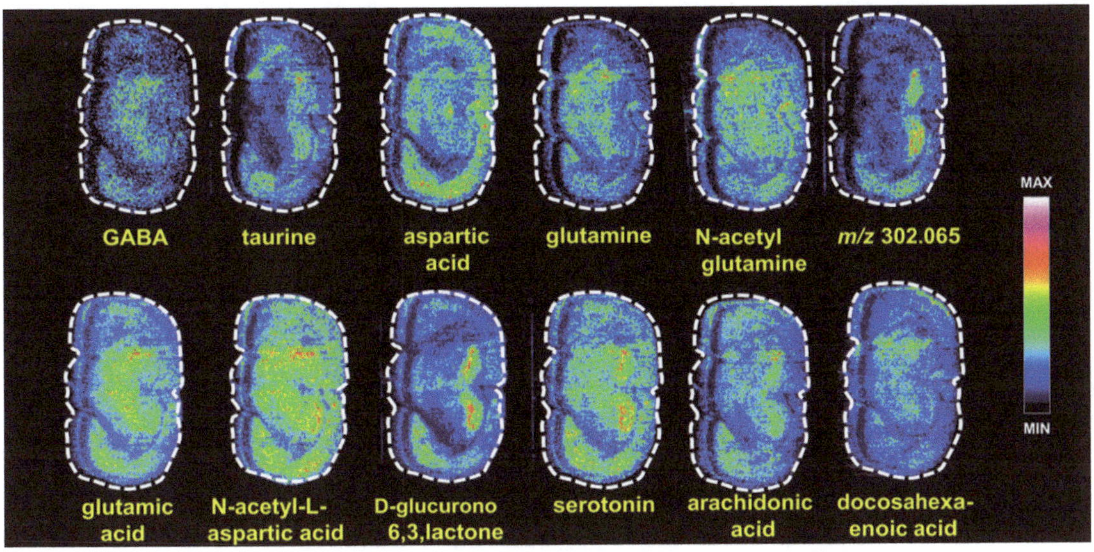

Fig. 3 Metabolite imaging of rodent brain section using DESI

5. Spatial correlation between detected metabolites and lipids was explored using analysis based on the Pearson product-moment correlation coefficient and hierarchical clustering analysis based on processed ions. Small molecules such as amino acids (e.g., taurine, glutamine, arachidonic acid) and neurotransmitters (e.g., GABA, serotonin) were simultaneously detected along with lipids (e.g., phosphatidylcholine, lysophosphatidylcholine) as shown in Fig. 3. See note on molecular annotation.

3.4 Collision Cross Sections Calculations

Metabolite annotation from mass spectrometry data remains one of the most challenging steps in any MSI workflow due to the biochemical complexity of the samples. In recent years, rotationally averaged collision cross sections (CCS) values derived from ion mobility MS have been used to support metabolite annotation. Many peer-reviewed journals and academic websites have publically available CCS values measured from standards or values derived from de novo calculated values using models. CCS values can be automatically calculated from drift-time using informatics products such as Progenesis® QI [7, 8]. For DESI HDMS imaging experiments, CCS values can be calculated manually from drift-times. A workflow is shown in Fig. 4.

1. Obtain a list of accurate mass *m/z* and drift bins from DESI-MS Imaging data with ion mobility (HDMS Imaging) from peak list in HDI.

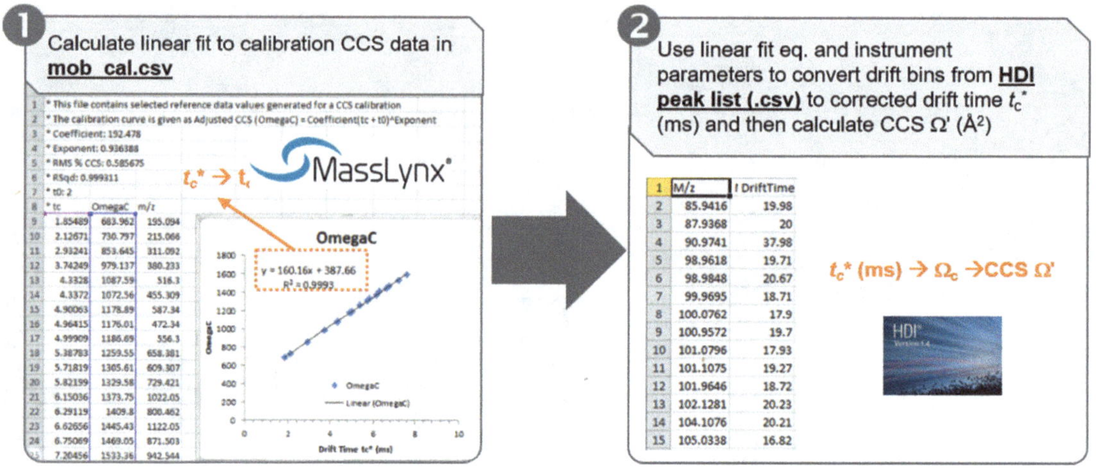

Fig. 4 Calculating CCS value in DESI imaging experiments

Table 1
Collision cross section (CCS) value obtained from DESI analysis

	Caffeine	Sulfaguanidine	Val-Tyr-Val	Verapamil	Terfenadine	Reserpine	LeuEnK
Accurate mass	195.0883	215.0598	380.2179	455.2919	472.3223	609.2819	556.2773
Theoretical mass	195.0882	215.0603	380.2185	455.2910	472.3216	609.2812	556.2771
Mass error (ppm)	**0.52**	**−2.32**	**−1.58**	**1.98**	**1.48**	**1.15**	**0.36**
DESI exp CCS (A^2)	136	144.6	191	208.3	227.1	249.4	227.8
Ref CCS value (A^2)	138.2	146.8	191.7	208.8	228.7	252.3	229.8
% CCS difference	**−1.51**	**−1.52**	**−0.28**	**−0.14**	**−0.61**	**−1.08**	**−0.8**

2. Obtain EDC Delay Coefficient (noted as C) and ADC pusher Frequency in μs (v_{ADC} = ADC pusher frequency) in _extern. Inf found within the MassLynx RAW folder.
3. Use the equation shown on the Fig. 4, **step 1** with the linear fit equations from a mob_cal.csv file found in the MassLynx RAW folder.
4. Using the equation shown on the Fig. 4, **step 2** convert the drift bins to corrected CCS in a spreadsheet such as Microsoft Excel.
5. Combining accurate mass measurement with CCS provides two orthogonal characteristics for annotating the cellular metabolites, thus increasing confidence in their identification and reducing the number of false positives [9]. A series of cellular metabolite standards were spotted onto a well plate. The CCS values calculated from the standards from their drift times to verify the accuracy of the CCS measurements are shown in Table 1 (*see* **Note 7**).

4 Notes

1. Label the orientation of the tissue on the mold before freezing.
2. Leaving fresh tissues sections in a desiccator too long will result in flaking due to excessive drying and may result in analyte delocalization.
3. Load excess volume of spray solvent to last more than total acquisition time of DESI imaging run.
4. Although mass spectrometer safety interlock automatically turns off the high voltage while adjusting emitter protrusion with a wrench, users are recommended to put the instrument on standby as a safety precaution.
5. Refer to HDI help files for detailed instructions regarding steps for setting up a DESI imaging experiment.
6. Note: Scan speed and spacing between lines (i.e., the total number of lines) will affect the total time of MS imaging. The true pixel size is slightly larger than the impacted area due to spreading of spray.
7. Molecular identification was aided using high mass accuracy (low PPM) database searches against publically curated databases, such as LipidMaps and Human Metabolome (HMDB). In addition to the accurate mass and high-fidelity isotopic distribution, collisional cross sections (CCS) or drift time data obtained during ion mobility separation was used to improve confidence in detected molecules.

5 Conclusion

DESI-MS imaging can be successfully applied to the tissue-level imaging. DESI imaging has not been demonstrated for a single mammalian cell or cell-by-cell imaging of metabolites in a most mammalian system. DESI HDMS imaging can depict intensity map of *m/z* peaks annotated as cellular metabolites on the tissue. Ion mobility separation prior to the MS was leveraged to increase coverage of the cellular metabolite species from tissue enhancing the detection of metabolite ions by discriminating them from electrospray background species, as well as improving identification of molecules by matching collision cross sections (CCS). DESI HDMS holds promise for cellular metabolism study and, with improvement in DESI technology, we can anticipate doing near single-cell level MS imaging with DESI.

References

1. Takáts Z, Wiseman JM, Gologan B, Cooks RG (2004) Mass spectrometry sampling under ambient conditions with desorption electrospray ionization. Science 306(5695):471–473
2. Watrous J, Hendricks N, Meehan M, Dorrestein PC (2010) Capturing bacterial metabolic exchange using thin film desorption electrospray ionization-imaging mass spectrometry. Anal Chem 82(5):1598–1600
3. Williams JP, Bugarcic T, Habtemariam A, Giles K, Campuzano I, Rodger PM, Sadler PJ (2009) Isomer separation and gas-phase configurations of organoruthenium anticancer complexes: ion mobility mass spectrometry and modeling. J Am Soc Mass Spectrom 20 (6):1119–1122
4. Fenn LS, Kliman M, Mahsut A, Zhao SR, McLean JA (2009) Characterizing ion mobility-mass spectrometry conformation space for the analysis of complex biological samples. Anal Bioanal Chem 394(1):235–244
5. Bush MF, Campuzano ID, Robinson CV (2012) Ion mobility mass spectrometry of peptide ions: effects of drift gas and calibration strategies. Anal Chem 84(16):7124–7130
6. Campuzano I, Bush MF, Robinson CV, Beaumont C, Richardson K, Kim H, Kim HI (2012) Structural characterization of drug-like compounds by ion mobility mass spectrometry: comparison of theoretical and experimentally derived nitrogen collision cross sections. Anal Chem 84(2):1026–1033
7. Paglia G, Menikarachchi L, Langridge J, Astarita G (2014) Travelling-wave ion mobility-MS in metabolomics: workflows and bioinformatic tools. In: Rudaz S (ed) Identification and data processing methods in metabolomics. Future Medicine, London
8. Paglia G, Williams JP, Menikarachchi L, Thompson JW, Tyldesley-Worster R, Halldorsson S, Rolfsson O, Moseley A, Grant D, Langridge J, Palsson BO, Astarita G (2014) Ion mobility derived collision cross sections to support metabolomics applications. Anal Chem 86 (8):3985–3993
9. Paglia G, Williams JP, Menikarachchi L, Thompson JW, Tyldesley-Worster R, Halldórsson S, Rolfsson O, Moseley A, Grant D, Langridge J, Palsson BO, Astarita G (2014) Ion mobility-derived collision cross-sections to support metabolomics applications. Anal Chem 86 (8):3985–3993

Chapter 15

Open-Source Software Tools, Databases, and Resources for Single-Cell and Single-Cell-Type Metabolomics

Biswapriya B. Misra

Abstract

In this age of –omics data-guided big data revolution, metabolomics has received significant attention as compared to genomics, transcriptomics, and proteomics for its proximity to the phenotype, the promises it makes and the challenges it throws. Although metabolomes of entire organisms, organs, biofluids, and tissues are of immense interest, a cell-specific resolution is deemed critical for biomedical applications where a granular understanding of cellular metabolism at cell-type and subcellular resolution is desirable. Mass spectrometry (MS) is a versatile technique that is used to analyze a broad range of compounds from different species and cell-types, with high accuracy, resolution, sensitivity, selectivity, and fast data acquisition speeds. With recent advances in MS and spectroscopy-based platforms, the research community is able to generate high-throughput data sets from single cells. However, it is challenging to handle, store, process, analyze, and interpret data in a routine manner. In this treatise, I present a workflow of metabolomics data generation from single cells and single-cell types to their analysis, visualization, and interpretation for obtaining biological insights.

Key words Software, Tool, Database, Mass spectrometry, Metabolomics, –Omics, Web server, Data, Pathway, Network, Analysis, Statistical, Computational, Single cell, Single-cell type, Microbial, Plant, Animal, Cell

Abbreviations

CE	Capillary electrophoresis
DB	Database
DESI-MS	Desorption ionization mass spectrometry
GC	Gas chromatography
GUI	Graphical user interface
HRMS	High-resolution mass spectrometry
HRMS/MS	High-resolution tandem mass spectrometry
KEGG	Kyoto encyclopedia of genes and genomes
LAESI-MS	Laser ablation electrospray ionization mass spectrometry
LC	Liquid chromatography
MS	Mass spectrometry
MS/MS	Tandem mass spectrometry

Bindesh Shrestha (ed.), *Single Cell Metabolism: Methods and Protocols*, Methods in Molecular Biology, vol. 2064, https://doi.org/10.1007/978-1-4939-9831-9_15,

NMR	Nuclear magnetic resonance
PCA	Principal component analysis
PLS-DA	Partial least square -discriminant analysis
QC	Quality control
QqQ	Triple quadruple
Q-ToF	Hybrid quadrupole orthogonal time-of-flight
R	R-programming language for statistical computing
ToF-MS	Time-of-flight mass spectrometry
UPLC	Ultra performance liquid chromatography
XCMS	Various forms (X) of chromatography mass spectrometry

1 Introduction

In this era of big data-driven life sciences / biomedical research and –omics (genomics, transcriptomics, proteomics, metabolomics, phenomics, etc.), enormous volumes of data in the scale of petabytes or higher are being generated on a daily basis. Although most of these studies focus on the traditional bulk level, that is, from whole organisms, systemic biofluids, intact organs, and tissues, the need for understanding the cellular status of individual transcripts, proteins, and metabolites remains challenging. To this end, single-cell –omics and single-cell-type –omics offer advantages in providing cell-specific information content that is lost in bulk –omics approaches. Furthermore, out of all the –omics, metabolomics takes the center stage in central dogma of life by connecting the genotype with phenotype, for its ability to provide a proxy output to the much stable genotype. Hence, numerous technological and analytical advances have been made in single-cell and single-cell-type metabolomics areas in terms of cell types investigated and platforms used [1, 2]. The number of samples processed through these high-throughput comparative "single-cell –omics" studies are increasing rapidly due to declining experimental costs, analytical advances, smart strategies, and hence confer advantages over the so-called traditional, or bulk, –omics datasets, Fig. 2 depicting workflow for mass-spectrometry-based data acquisition, and finally Fig. 3 with the list of open-source software tools, packages, databases, webservers, etc. (*See* **Note 1**).

Coupled with other single-cell –omics approaches (such as single-cell genomics, single-cell transcriptomics, and single-cell proteomics), the information content of single-cell metabolomics are going to be more informative for translational approaches—as it would reflect the function of a singular unit of life (i.e., cell) in the context of wellness, disease status among others. Of course, there is a necessity to combine single-cell omics datasets with positional information (i.e., spatial resolution) to meaningfully address biological questions of importance, ranging from agriculture,

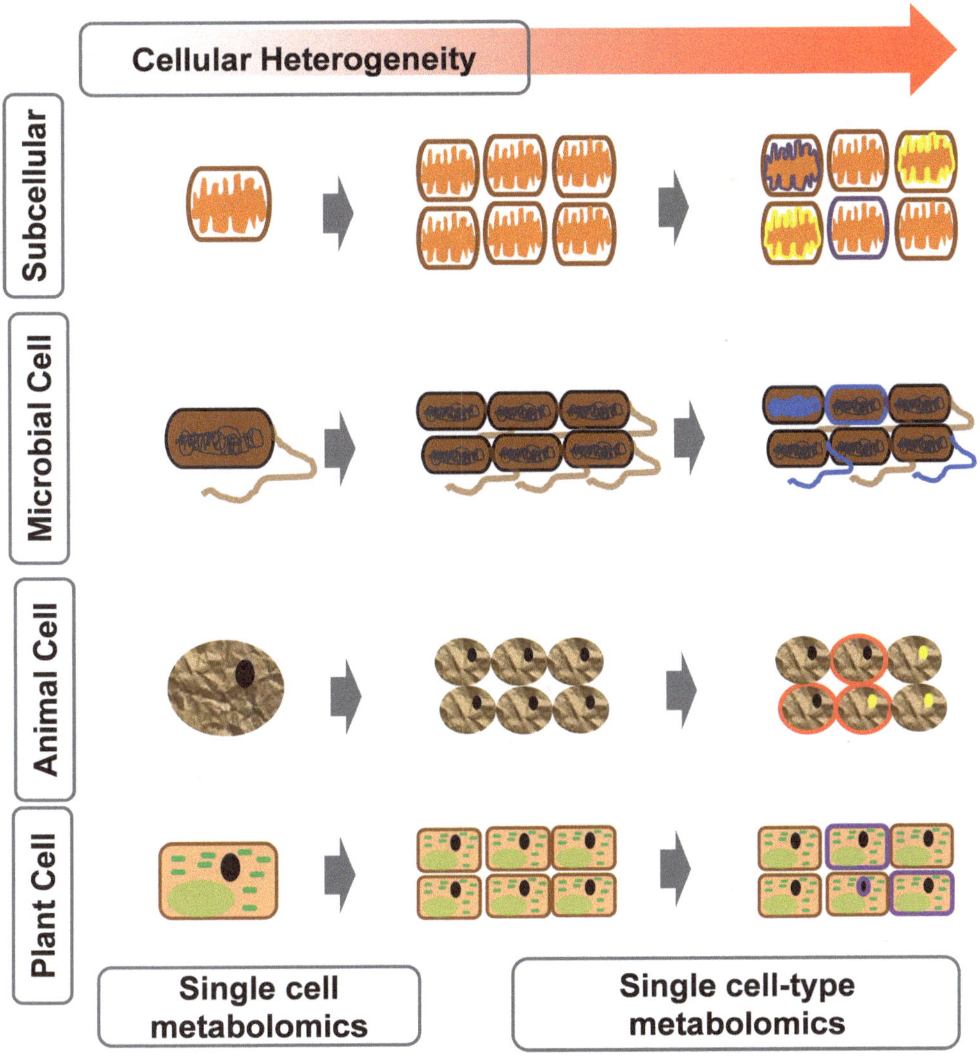

Fig. 1 Schematic representation of single-cell and single-cell-type metabolomics at subcellular resolution, and in animal, plant, and microbial cells. Challenges increase with increased cellular complexity (division of labor and multicellularity) and heterogeneity (from spatiotemporal diversity among others) (*see* **Notes 2, 3**)

medicine, and healthcare to space biology. Toward this end, for instance, *the Human Protein Atlas* portal (www.proteinatlas.org) for proteomics [3] and the *GTEx* portal (http://www.gtexportal.org/home/) for genomics [4] are commendable, given that a similar database for human metabolites with cell-type specificity is clearly lacking. Furthermore, multiomics studies of single cells have shown tremendous potentials and applications in basic and biomedical research [5]. Although single-cell sequencing-based technologies have taken off [6] usually mass spectrometry (MS)-based –omics such as single-cell proteomics and single-cell metabolomics have not reached there yet for numerous technological and

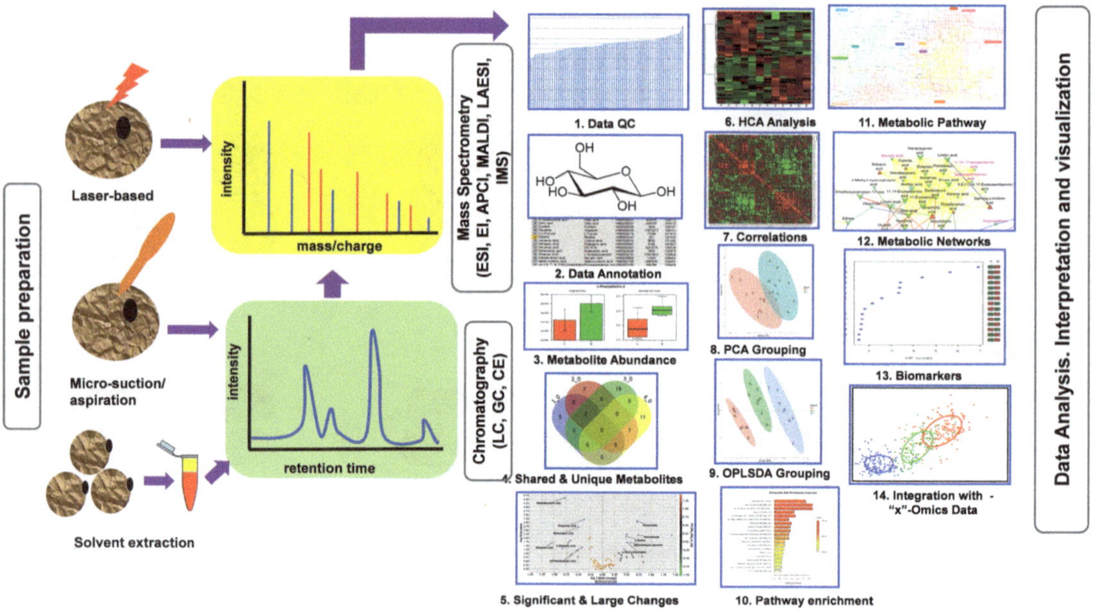

Fig. 2 Workflow for mass spectrometry-based data acquisition and analysis for single-cell and single-cell-type metabolomics. (**a**) Sample preparation from a single cell can be using laser-based desorption or ablation methods, or microsuction, aspiration, or capillary mediated. Single-cell-type metabolomics data is usually acquired from a homogenous cell population using solvent extraction (*see* **Notes 4, 5**). Once sample preparation is done, either (**b**) chromatography (LC, GC, CE)-based (**c**) mass spectrometry (MS) methods (ESI, EI, APCI, MALDI, LAESI) or MS alone is performed (*see* **Note 6**). Once the data is acquired, data preprocessing would precede annotation, statistical analysis (univariate and multivariate), pathway and network analysis, followed by biomarker discovery and/or data integration with other -omics datasets

analytical limitations in their adaptability for routine scientific discoveries. Although merging of single-cell biology and other single-cell omics data look tremendously beneficial, as the temporal and spatial information, robust single-cell techniques would enable investigators to test omics-derived hypothesis with ease.

Resources such as "awesome-single-cell" (GitHub, https://github.com/seandavi/awesome-single-cell) enlists software packages for single-cell –omics (mostly dedicated to RNAseq/transcriptomics and genomics data analysis). For single-cell RNA-seq data analysis, it lists >50 software packages, most of which are R-statistical programming language or Python based. The author has described all available tools, software, resources, and databases for bulk metabolomics data analysis over the past few years [7–11]. Although it is heartening to see that all these existing tools serve the genomics community only which is based on high-throughput sequencing efforts and bulk metabolomics data analysis, it is unfortunate that no such single-cell proteomics or metabolomics resource exists at all—given both of them rely on MS-based platforms for data generation in sharp contrast to genomics platforms, that are sequencing based. Although the single-cell

Fig. 3 List of open-source software tools, packages, databases, webservers, and other resources available for the analysis of single-cell and single-cell-type metabolomics data. Informatics solutions are categorized here for convenience puposes as (1) Data formatting and conversion, (2) Preprocessing, (3) Metabolite annotations/identification, (4) Statistical analysis, (5) Naming and ID conversions, (6) Data visualization (i.e., networks and pathways), (7) Multifunctional tools, (8) Metadata, data handling and storage/ archiving, and (9) Data integration. Note: Many of the above-described tools fall into multiple categories as well and perform multiple functions

genomics and transcriptomics community has attempted to address the challenges faced in a very timely manner, little to no such efforts exist in the metabolomics community. Data processing for any metabolomics datasets are comparable, although single-cell and single-cell-type metabolomics data poses its own challenges such as dealing with cellular heterogeneity, false positives, cell-type deconvolution among others. NMR spectroscopy-based tools for metabolomics data analysis is another huge area of interest and discussion, and I have refrained from addressing it for lack of expertise and little advances in the area. In this chapter, I demonstrate a workflow for single-cell and single-cell-type metabolomics data generation using MS, to data processing, analysis, and interpretation in lieu of the existent tools and what needs to be done. Fig. 1 displays a schematic presentation of single-cell and single-cell-type metabolomics at subcellular resolution, and in animal, plant, and microbial cells.

Just like the steps involved in any other –omics technology, for single-cell metabolomics the steps can be delineated as (1) *cell or cell-type identification*, (2) *sample preparation* (cell or cell-type isolation, extraction of the analytes), (3) *sample measurement* [i.e., analyte separation using chromatography and analyte detection using either mass spectrometry (MS) or spectroscopy (such as NMR, IR, Raman)], and (4) *data analysis* (data preprocessing, annotation of the metabolites, statistical analysis, analysis of pathways/and networks, data visualization, and biological interpretation of the data). A quick guide to performing a metabolomics study, starting from experimental design to data preprocessing, statistical analysis, data visualization, biological interpretation, to data archiving for bulk metabolomics analysis is provided elsewhere [8]. Moreover, bioinformatic tools, software, and databases for metabolomics research are constantly being accumulated at online resources such as *OMICtools* (http://omictools.com/) [12], Fiehn Lab resources (http://fiehnlab.ucdavis.edu/), the International Metabolomics Society's resource pages (http://metabolomicssociety.org/resources/metabolomics-databases), *ms-utils.org* (http://www.ms-utils.org/wiki/pmwiki.php/Main/SoftwareList), Pacific Northwest National Laboratory (PNNL)'s Integrative Omics open-source software collections (https://omics.pnl.gov/software/), etc. Keeping in mind the readers, I introduce available resources for the first three steps that are critical prior to the metabolomics data interpretation.

1.1 Cell (Cell-Type) Identification

Classical biological approaches such as anatomy, histology, and microscopy allow defining the cell (and cell-type) to be studied. Much is governed by the biological questions at hand, which drive the manipulation of a given cell-type. In addition, single-cell analysis (SCA) methods and in vitro cultures of plant, animals, and microbial cells have allowed analysis of hundreds of cell types. Out of bottom-up and top-down approaches, that is, from established cell-type to metabolomics, data acquisition has been feasible, while data deconvolution to cell-type prediction has been virtually non-existent. This partly limited due to less number of studies and coverage of metabolites from a single cell-type. Thus, target cell-type detection and experimental design is of paramount importance.

1.2 Sample Preparation

Examples of single-cell isolation and trapping procedures include those that rely on established hydrodynamic, optical, magnetic, or dielectrophoretic, and acoustic traps [13]. For single-cell, single-cell-type, and subcellular metabolomics studies, micropipetting/micromanipulation/microsuction, laser capture microdissection (LCM), enzymatically enriched preparations, in vitro cultures (plant and algal suspension cells, bacterial cultures, and animal cell lines) adherent or nonadherent, and differential (e.g., sucrose gradient) centrifugation techniques for subcellular resolution have been successfully used over the years. However, recent approaches in live-cell metabolomics have circumvented several limitations as well. For instance, a real-time metabolome profiling achieved by direct injection of living bacteria, yeast, and mammalian cells in a high-resolution mass spectrometer (HRMS), resulted in the automated monitoring of about 300 compounds in 15–30 s cycles in a matter of few hours [14].

Metabolomics data acquired is usually platform dependent. Hence, the choice to use or not use a given chromatography [liquid chromatography (LC), gas chromatography (GC), capillary electrophoresis (CE), etc.] at the front end of MS depends on the compounds of interest and experimental strategies. Other MS-based technologies available are laser ablation electrospray ionization (LAESI)-MS, desorption electrospray ionization (DESI-MS), matrix-assisted laser desorption/ionization (MALDI- MS), direct injection (DI-MS), and several spectroscopy-based platforms such as nuclear magnetic resonance (NMR), Raman where chromatography is not a prerequisite. After preparation of a single cell, single cell types, or subcellular fractions, the next step focuses on the extraction of analytes, which is done off site chemically (usually with organic solvents and/aqueous mixtures, afterwards needing a chemical derivatization or not), and/or mechanically [in situ using aspiration (microsucking, micropipetting)] methods. For instance, in a microfluidic device-mediated single-cell analysis, or a lab-on-chip analysis, the detection methods can be fluorescence detection, amperometric detection, or MS-based detection [13]. Depending on whether the study is a single-cell metabolomics (chromatography not preferred) or a single-cell-type or subcellular metabolomics study (both cases where enrichment is feasible), chromatography (LC, GC, and CE) is typically performed. With due consideration to quality control (QC) measures [15], such as the usage of suitable positive and negative controls, internal standards, retention time (index) markers, and means to include QC samples intermittently, quality assurance (QA) of the analytical system suitability, ways to avoid batch, unwanted sources of variations, and technical variability, the metabolomics data are acquired for further downstream processing and statistical analysis. For a more detailed insight into GC-MS-based bulk metabolomics data acquisition and analysis, the readers are suggested to *see* Lisec et al. publication as referenced

[16]. For a very detailed workflow for untargeted LC-MS-based data generation and analysis using XCMS and annotation at METLIN, the readers are advised to consult Zu et al. as referenced [17].

1.3 Sample Measurement

Data acquisition itself can be really challenging and is a matter of choice for the user—whether to perform untargeted MS-based metabolomics that can also leave the user with a lot of unknown metabolite features (i.e., defined as a pair of *m/z* and retention time) with difficulties in their annotation as well as distinguishing them from the noise, or to perform targeted metabolomics experiments, though sensitive, but depending on the spatiotemporal constraints and the breadth of chemical space/coverage of the instrument a smaller subset might be captured depending on the analytical limitations of the platform. Moreover, owing to obvious challenges in isotope-labeled approaches in single-cell metabolomics (exceptions—feeding experiments or full-scale labeling in cultured cells), most data acquisitions rely on relative abundances (based on peak height ion counts or area under the peaks), as technical reproduction of MS measurements neither characterizes the whole experiment from cell cultivation to producing biological data nor says anything about systematic errors in the overall measurement procedure [18]. In addition to approaches such as heuristically predicting the fragmentation patters in silico and to compare these to experimental tandem mass spectrometry (MS/MS), spectra, the availability of Fourier Transform Ion Cyclotron Resonance (FTICR) or Orbitrap^{TM} Fourier transform MS methods have provided several exciting alternatives to generate accurate mass (AM) spectra of MS peaks for their structural elucidation and confident identification via sub-structure prediction among other in silico approaches.

Mass spectrometry imaging (MSI) is a technology amenable for spatial analytics that has experienced a significant interest in recent years. Data generated using various ionization sources for MS-based metabolomics data acquisition strategies such as matrix-assisted laser desorption/ionization (MALDI), SIMS, and DESI are in the scale of >1 TB/day [19]. For spatially resolved single-cell metabolomics applications through MSI, the use of secondary ion mass spectrometry (SIMS) (resolution of below 1 μm), MALDI with UV/IR lasers, laser ablation electrospray ionization (LAESI) with IR lasers have been shown to be immensely useful in diverse applications [20] in plants to animals and clinical samples. MSI by LAESI-MS allows mapping of metabolite distributions in one to handful of cells in untreated (or live) samples. Further, LAESI-MS can be used to analyze adherent cells directly, while still attached to a culture surface, where incorporating an objective with a high numerical aperture, spot sizes of 10–20 μm are readily achieved for analysis of only five adherent cells in culture conditions [21]. Ion mobility-MS (IMS) is another growing area of metabolomics research, where ion mobility-derived collision cross-sections (CCSs) aid in metabolite identification [22] as an orthogonal measure to compound identification in addition to tandem MS. As single-cell metabolomics, either through IMS or chromatography hyphenated or direct injection MS (DIMS), is still evolving platforms, most laboratories have tend to use their own custom-made software tools or have heavily relied on the commercial software sold by the vendors of those MS platforms. However, new open-source tools are being populated in the MS-data analytics space on a daily basis.

2 Materials

2.1 Equipment

2.1.1 Software

All robust and popular web browsers that support HTML5 (HyperText Markup Language) and JavaScript, such as the latest Google Chrome or Mozilla Firefox are required for best performance during metabolomics data analysis for several web-based applications. Most software described and the R-packages run on a 32-bit or 64-bit version of Microsoft Windows operating system (OS) while a handful are amenable to Linux and Mac OS. The processing time needed for the steps of analytical workflow depends on the size of the data, the nature of the platforms used to generate the data, the computer's memory and processing speed, and user's computational expertise level among other factors. Majority of the open-source tools described in this chapter are either web-server based, are standalone, or are mostly R-programming based.

GitHub is a web-based Git repository hosting service. It offers all of the distributed version control and source code management (SCM) functionality of Git, that serves users to host, distribute, and use open-source software projects. The readers are suggested to read Perez-Riverol et al. [23], for more detailed account of GitHub.

R (https://www.r-project.org/) is "GNU S," a freely available language and environment for statistical computing and graphics which provides a wide variety of statistical and graphical techniques: linear and nonlinear modeling, statistical tests, time series analysis, classification, and clustering among others. Further, *CRAN* (The Comprehensive R Archive Network, https://cran.r-project.org/web/packages/available_packages_by_date.html) is a network of FTP and web servers around the world that store identical, up-to-date, versions of code and documentation for R. Moreover, the integrated development environment (IDE) for R, *RStudio* is immensely useful (https://www.rstudio.com/products/rstudio/). *Bioconductor* (https://www.bioconductor.org/) is primarily based on the statistical programming language R, but does contain contributions in other programming languages, and is a free, open-source and open-development software project for the analysis and comprehension of –omics and other biology-based data generated using wet-laboratory experiments.

Galaxy is an open-source, web-based platform for *accessible*, *reproducible*, and *transparent* computational biomedical research. Galaxy is immensely useful in analysis of the individual and multiomics data sets [24]. A list of tools in the Galaxy Tool Shed can be found here: https://toolshed.g2.bx.psu.edu/ (search—Repositories in Category Metabolomics) for the metabolomics community. Publicly hosted Galaxy development projects at academic institutions include http://galaxy.nbic.nl/ and http://galaxy.wur.nl/. Fig. 2 depicts a generalized workflow for mass-spectrometry-based data acquisition in single cell and single-cell-types.

2.1.2 Hardware

Depending largely on the size of the data, however, at least a 3.2 GHz CPU (Intel Core i5/i7), 32/64 GB or higher physical RAM with at least 8 GB free, and few 100 GB/ few TBs of free storage space, and a minimum of a 17-in. "dual" screen with a screen resolution of 1280 × 800 or higher are recommended. For some software described here in need parallel execution on multiple cores, in which case at least 2 GB of RAM is required per thread that is running in parallel. Access to an high performance computing (HPC) cluster is also highly desirable for large-scale datasets runnig into 1000s of files or more- for automation to multitasking, and would need experts to run programs using scripts/ command line intrefaces.

2.1.3 Data Files

Raw data files are typically obtained from a given analytical platform (i.e., GC-MS, LC-MS, MALDI-ToF-MS, CE-MS, LAESI-MS, etc.) by a user in their own laboratory set up (or, that of a core facility) and originate from diverse platforms. Most of the tools described in the following sections for analysis of metabolomics-scale data typically provide (within the package) or are downloadable as example datasets for trial runs.

Moreover, handful of public data archives and repositories such as MetaboLights [25], Metabolomics Workbench [26] and Global Natural Products Social Molecular Networking (GNPS)'s [27, 28] servers help archive raw datasets from studies where the researchers intend to submit the data. For instance, a typical MALDI-MS-based metabolomics data files can be downloaded from MetaboLights (http://www.ebi.ac.uk/metabolights/) (study ID MTBLS29, *Single-cell and bulk level metabolomic study of the glycolysis inhibition by 2-deoxy-D-glucose in Saccharomyces cerevisiae*), or GC-MS and LC-MS data (study ID, MTBLS175, *Metabolomics-derived insights into the manipulation of terpenoid synthesis in Centella asiatica cells by methyl-jasmonate*), or the very recent gold-standard dataset from 3D MALDI- and DESI-IMS platform that has datasets from murine pancreas, murine kidney, human oral squamous cell carcinoma, and interacting microbial colonies cultured in Petri dishes in ".imzML format" (study ID, MTBLS176, *Benchmark datasets for 3D MALDI- and DESI-Imaging Mass Spectrometry*). From the Metabolomics Workbench raw and metadata are available for studies on MDA-MB-231 cells (study ID, ST000043) and *Synechococcus* culture cells (study ID, ST000318), among a few hundred of other available datasets originating from a diverse array of studies. From GNPS, as an example, skin etabolomics datasets from 12 volunteer human beings (study ID, MSV000081582) with 4,594 files of about 470 GBs size are avialable for free download and reuse, among another ~1200 such studies availalble.

Dryad (https://datadryad.org/) is another generic international disciplinary repository of data underlying scientific and medical publications, which is a curated general-purpose repository that makes data discoverable, freely reusable, and citable. A decent number of studies have started depositing their metabolomics scale data into this repository as well.

2.1.4 Image Processing

A recent review, for instance, details all available open-source software solutions for evaluation and visualization of laser ablation inductively coupled plasma mass spectrometry imaging (LA-ICP-MSI) data [29]. In the following sections, I would introduce not only tools for image analysis (i.e., MSI-based analysis), but also tools amenable to all kinds of MS-based metabolomics datasets. One also needs image processing tools for the cells or cell types for generating suitable images and for eventual publication. *ImageJ* (https://imagej.nih.gov/ij/download.html) can be used on Mac OS X, Linux, and Windows platforms for image processing and experimental data visualization. *Inkscape* (https://inkscape.org/en/) is an open-source vector graphics software for creating illustrations, images, diagrams, and publication quality web graphics. By no means the tools listed here are exhaustive, and the readers are recommended to consult excellent reviews available on open-source libraries and frameworks for biological data visualization [27, 30].

2.1.5 Resources and Record Keeping for Reproducibility

In lieu of the laboratory information management system (LIMS) tools that are usually licensed, commercial, expensive, and not open sourced, for laboratory resource management and sample ID/record-keeping purposes, free tools such as Quartzy (focused on the organization of inventory, https://www.quartzy.com) or Labguru (designed for laboratory project management, http://www.labguru.com) [31] can be used for sample storage/ archiving information, retrieval, data storage, and record keeping, among others. These efforts would lead to FAIR (Findability, Accessibility, Interoperability, and Reusability) practices for scientific data management and stewardship [32], in addition to adhering to robust experimental designing strategies.

3 Methods

3.1 Data Formats

The acquired data from a typical MS platform are usually available in a format that is specific to a MS vendor and are amenable to downstream processing only using their commercial software solutions. However, data available in these diverse formats can be converted to formats that are vendor neutral, i.e., open-source tool compatible formats such as .mzXML, .mzML, .mzData, .netCDF (or, ANDI-MS), .mgf, and .csv among others which save the spectral data and the intensity data among other features (i.e., retention time) which are essential for compound identification and quantification. Towards this, ProteoWizard's [33] tool *msConvert* (http://proteowizard.sourceforge.net/tools.shtml) is capable in converting most vendor-specific commercial file formats to the aforementioned open-source formats. *OpenChrom* (https://openchrom.net/) [34] is another excellent open-source tool aiding in data format interconversions. *mMass* (http://www.mmass.org/) is another tool that not only helps in data conversion, but also works on multiple operating systems, supports all modern open formats like .mzML, .mzXML, .mzData, ASCII, and data preprocessing such as recalibration, smoothing, baseline correction, auto peak picking, and deconvolution among others [35]. *MALDIquantForeign*, an R -package, helps in reading (.txt, .csv, Bruker's .fid, Ciphergen XML, .mzXML, .mzML, .imzML, Analyze 7.5, .netCDF, mMass MSD) and writing (.txt, .csv, .mMass MSD, .mzML, .mzML) different file formats of MS data into/from MALDIquant objects.

3.2 Data Preprocessing

For data preprocessing, there are several metabolomics research community-wide adopted tools that are amenable to LC-MS-based data analysis such as *MZmine* 2 [36], which performs data processing steps such as peak detection, peak filtering, chromatogram building, chromatogram deconvolution, isotopic peak grouping, chromatogram alignment, gap filing, adduct search, and peak identification by multiple spectral and chemical databases. Other popular tools, including *XCMS Online* (https://xcmsonline.scripps.edu/) webserver [37–39], can process and analyze data generated from Agilent Technologies, Sciex, Bruker, Thermo Fisher Scientific, and Waters hardware, and are very interactive and produce nice visualizations such as the Cloud plots and provides an user-friendly web interface to the *XCMS* algorithm [40]. Other new robust tools include *Mass*$^{++}$ [41] (http://www.shimadzu.co.jp/aboutus/ms_r/masspp.html), *MS-DIAL* (http://fiehnlab.ucdavis.edu/projects/MS2Analyzer/) [42] among others. For LC-MS-based metabolomics data, *MAVEN* (Metabolomic Analysis and Visualization Engine) (http://genomics-pubs.princeton.edu/mzroll/index.php) performs data import; peak alignment across samples; identification of metabolites that differ strongly between biological conditions, quantitation, and visualization of isotope-labeling patterns; and export of tables of metabolite-specific peak intensities [43]. *GalaxyM* is a Galaxy-based data-processing workflow for DIMS and LC-MS-based metabolomics [44]. Fig. 3 provides a list of open-source software tools, packages, databases, webservers, etc.

For GC-MS-based metabolomics data analysis, one can use the freely available National institute of Standard Technology (NIST)'s Automated Mass Spectral Deconvolution and Identification System (*AMDIS*) [45] for deconvolution of spectra before annotation of the spectra are pursued using user-chosen library. Other open-source and popular tools to handle GC-MS-based metabolomics data include *OpenMS* [46], *MS-DIAL* [42], *MZmine 2* [36], *XCMSOnline* [39] among others. For interpretation of accurate tandem, mass (MS/MS) spectra *MS2Analyzer* [47] is also available. Specifically, for lipidomics in silico tandem mass datasets *LipidBlast* [48] is a useful software. *MetAlign* (http://www.wur.nl/nl/Expertises-Dienstverlening/Onderzoeksinstituten/rikilt/show/MetAlign.htm) performs data preprocessing such as automatic format conversions, accurate mass (AM) calculations, baseline corrections, peak-picking, saturation and mass-peak artifact filtering, as well as alignment of up to 1000 data sets—for both nominal and AM spectra obtained from LC-MS and GC-MS platforms [49]. *Decon2LS* (https://omics.pnl.gov/software/decontools-decon2ls) is an open-source software package for automated processing and visualization of high-resolution mass spectrometry data (HRMS) ESI-MS-derived metabolomics data [50].

For MALDI-MS and DESI-MS data processing, Matlab-based tool, *omniSpect*, transforms time-series mass spectra into MS images based on a probe spatial raster path and multivariate analysis [51]. *MALDIquant*, an R-package, is a complete analysis pipeline for MALDI-TOF data analysis that includes functionalities such as baseline subtraction methods such as morphological filters (TopHat) or the statistics-sensitive nonlinear iterative peak-clipping algorithm (SNIP), peak alignment using warping functions, handling of replicated measurements, as well as allowing spectra with different resolutions for quantitative data workflows. *Boston University Data Analysis* (BUDA) (http://www.bumc.bu.edu/ftms/buda/) is an open-source free software project to develop analysis software for Fourier Transform MS data analysis [52].

3.3 Metabolite Annotation

Once the processed data is available as a .csv (comma separated) or .txt (tab-delimited) file format containing the information on the metabolite features [(*m/z*, retention time, abundance (intensity counts or area under the peak information)], these files are further processed for metabolite annotation (dereplication) with or without statistical analysis.

For GC-EI-MS-based metabolomics datasets generated using GC-ToF-MS or GC-Q(QQ)-MS experiments, the *GOLM DB* (http://gmd.mpimp-golm.mpg.de/) [53] is a freely available and immensely rich resource for annotating compounds that are obtained post-derivatization in a gas phase (i.e., silylation). Although originally developed for plant metabolomics community, the metabolites are common to most biological systems and cover the basic cellular metabolism. For EI-MS data, the *NIST Isotope Calculator* (ISOFORM, Version 1.02) helps derive the calculated monoisotopic masses and visualize the isotope distribution patterns. SWGDRUG Mass Spectral Library (http://www.swgdrug.org/ms.htm) has compiled an EI mass spectral library from a variety of sources, containing drugs and drug-related compounds constituting ~2440 EI-MS spectra amenable for library searching.

For ESI-MS-based data generated typically, using LC-MS, UPLC-MS, and CE-MS platforms, there are numerous mass spectral databases available such as *METLIN*, Human Metabolome Database (*HMDB 3.0*) [54], Kyoto Encyclopedia of Genes and Genomes (*KEGG*) [55], and *MassBank* [56], among others. *Mass Bank* (www.massbank.jp) is very useful for annotation of LC, GC, and CE-MS generated data sets as well as MALDI-ToF, QQQ, Q-ToF, EI, ESI, and APCI—options to choose from. The *mzCloud* (https://www.mzcloud.org/) follows a spectral-tree-based approach toward compound annotation [57]. *ReSpect DB* (http://spectra.psc.riken.jp/) [58] is useful in annotation of tandem MS (MS/MS) spectra.

For LC-MS-based-specific applications oriented toward natural products discovery and microbial metabolomics investigations, resources such as *Global Natural Products Social Molecular Networking* (GNPS; http://gnps.ucsd.edu) server is a very useful tool for spectral data processing, unknown annotation, network-scale data visualization, and sharing of the results (i.e., spectra, molecular networks) and raw datasets. For plant-based natural products annotations, KNApSAcK (http://kanaya.naist.jp/KNApSAcK/) [59] is a very useful resource.

Metabolic In silico Network Expansions (*MINEs*), is a database (http://minedatabase.mcs.anl.gov/#/home) that is an extension of known metabolite databases to include molecules that have not been observed, but are likely to occur based on known metabolites and common biochemical reactions [60], currently has over 571,000 compounds, of which 93% are not present in the PubChem database. For annotations of both LC-MS and GC-MS-generated spectra, *Mass Bank of America (MoNA)* is a newer resource (http://mona.fiehnlab.ucdavis.edu/).

MycompoundID.org (http://www.mycompoundid.org/mycompoundid_IsoMS/) is a web-based resource for identification of compounds of interest based on chemical properties of a molecule, such as AM and fragment ion spectral pattern generated by MS [61]. CFM-ID (http://cfmid.wishartlab.com/) provides a method for accurately and efficiently identifying metabolites in a given mass spectra generated by electrospray tandem mass spectrometry (ESI-MS/MS) [62].

3.4 Statistical Analysis

Prior to any further in-depth statistical analysis, the data is corrected for unwanted sources of variations (other than biological question in hand) such as batch variations, carryovers, and technical variabilities from handling and sample preparation among others by various strategies. These include corrective measures and data preprocessing strategies such as normalization, transformation, and scaling. For instance, to correct for batch variations (evident in large studies where samples are run over weeks or months across several lots) *limma*, available as an R-package, *normalyzer* (http://quantitativeproteomics.org/normalyzer) as a standalone tool/webserver [63], popular methods such as *LOESS* [64] normalization are carried out on the data to account for technical and batch variability, among others. Diverse generic statistical tools applied to other bulk metabolomics research community can be used for single-cell and single-cell-type metabolomics data analysis as well. Statistical analyses are typically performed using the statistical software *R* (Version 3.5 and above, R Development Core Team 2014, http://www.R-project.org), where the number of packages that are available for metabolomics-scale statistical data processing are growing on a monthly basis. Other open-source tools available include those from *Galaxy*, or those amenable to *Python*, *C++*, and *Java* platforms. Depending on the study design, univariate analysis such as t-tests, Analysis of Variance (ANOVA), or multivariate analyses such as unsupervised principal components analysis (PCA) and orthogonal partial least-squares discriminant analysis (OPLS-DA), sparse PCA (sPCA) can be performed to show the clustering of samples, and discrimination between groups (i.e., control vs. disease etc.) (*see* **Note 7**). More advanced analyses such correspondence analysis (CA) and Multiple Correspondence Analysis (MCA) are also performed using available R-packages with the same names or the package *FactoMineR* [65].

BioStatFlow version 2.7.7 (http://biostatflow.org/), developed by National Institute of Agronomic Research, France (INRA, http://www.inra.fr/), is an extremely user friendly web tool for analyses of metabolomics data. *DeviumWeb* (Dynamic MultivariatE Data Analysis and VIsUalization PlatforM) is an R-package available for data analysis and visualization software for both univariate and multivariate data [66]. *DeviumWeb* offers a variety of tools for metabolomic data analysis, including statistical and power analysis, data normalization, clustering, PCA, machine learning methods, and biochemical pathway enrichment analysis and visualization tools (*see* **Note 8**).

3.5 Naming and ID Conversions

Once the compound annotation have generated the names associated with the metabolite features or have assign them as metabolites, the common or chemical names derived from the searched database are converted to their more recognizable and uniform identifiers (IDs) for the ease of data sharing and archiving purposes. These IDs include *KEGG* IDs, *HMDB* IDs, *PubChem* (https://pubchem.ncbi.nlm.nih.gov/) IDs, *LipidMaps* (http://www.lipidmaps.org/) IDs, *InChiKey* (http://www.inchi-trust.org/) IDs, Chemical Abstracts Service (CAS) (http://www.cas.org/content/chemical-substances) IDs, Chemical Entities of Biological Interest (ChEBI) (https://www.ebi.ac.uk/chebi/) IDs, and *ChemSpider* (http://www.chemspider.com/) IDs among others. Such ID conversions can be performed one by one or in batches using functionalities available at MetaboAnalyst 4.0 [67], Fiehn Laboratory hosted Chemical Translation Service (CTS) (http://cts.fiehnlab.ucdavis.edu/conversion/batch), or *MBROLE* webserver [68]. *BridgeDb* (http://www.bridgedb.org/) is a framework for finding and mapping equivalent database identifiers and is both a framework, provides live services, and identifier mapping files for genes, proteins, and metabolites [69].

3.6 Data Visualization (Network and Pathway Analysis)

Once the user obtains the metabolite IDs, the next stage of analysis involves analysis of metabolites in the context of pathway and network analysis. Pathway analysis helps user map the metabolites to known biochemical pathways available across diverse pathway databases. For instance, using *KEGG Pathway Mapper* tool (http://www.genome.jp/kegg/mapper.html), the metabolites can be mapped onto the metabolic maps or individual metabolic pathways available at the KEGG database. Pathway-based visualizations are also possible at Small molecule pathway database (*SMPDB 2.0*) (http://www.pathvisio.org/) [70], *BioCyc* (http://biocyc.org/overviewsWeb/celOv.shtml) [71], *Pathvisio* (http://www.pathvisio.org/) [72], and *Pathomx* (http://pathomx.org/) [73]. *MarVis-Pathway* tool (http://marvis.gobics.de/) is another tool that allows users to identify metabolites by annotation of pathways based on mapping of data set features by ID, name, and AM, and also from adduct and isotope correction of MS data [74]. *WikiPathways* (http://www.wikipathways.org) is an open, collaborative platform for capturing and disseminating models of biological pathways for data visualization and analysis.

Dataset enrichments (or overrepresentation analysis) in terms of their relevance to a biological phenomenon, disease, tissue, or cell specificity are performed at Metabolite set enrichment analysis (*MSEA*) (http://www.msea.ca/MSEA/) and Metabolomics pathway analysis (*MetPA*) (http://metpa.metabolomics.ca/MetPA/), both modules now integrated with the *MetaboAnalyst* webserver. Unlike the genomics community, for enrichment of datasets where Gene Ontology (GO)-based approaches are popular, for metabolomics datasets metabolite ontology (MO) are still in the development phase and is awaiting release.

For a network-oriented view of the data (thus bypassing the classical view of pathways and to help facilitate unbiased biochemical connectivity guided novel biological discovery), the users have the option to use *STITCH 5.0* database [75]. *MetExplore* (http://metexplore.toulouse.inra.fr/joomla3/index.php) allows integration of untargeted metabolomics data into context of genome-scale modeling datasets [76]. *MetaNET* (http://metanet.osdd.net/) is an open-source, platform-independent, and web-accessible resource for metabolic network analysis through flux balance and variability, chemical species participation, cycles, and extreme paths identification [77]. *Reactome* [78] is another webserver where both network and pathway perspective of the metabolites are taken into context for analysis of metabolomics datasets. Alternatively, these generated networks can be exported to interactive open-source standalone tools like *Cytoscape* [79, 80] for taking advantage of a huge number of applications available for further refined analysis for enrichment, pathway analysis, and visualization of networks. *FunRich* is an open access, standalone functional enrichment and network analysis tool that performs functional enrichment analysis on background taxonomic and custom-made databases that are integrated from heterogeneous omics and metabolomics datasets [81]. For 3D visualization of biological surfaces (such as skin, leaf, and appendages) for "molecular cartography," the ion intensities can be mapped onto the 3D models using *ili*, a spatial visualization tool (https://ili-toolbox.github.io/).

3.7 Multifunctional Tools

On contrary to all the tools discussed above, which furnish a single or a specific set of task, there exists several tools which cater to the needs of the metabolomics research community by using raw (or preprocessed) data as inputs and providing excellent visualizations as the final products.

MetaboAnalyst [82] (www.metaboanalyst.ca) remains one of the most popular webserver for conducting data processing, both univariate and multivariate statistical analysis and data visualization. In addition, *XCMSOnline* (https://xcmsonline.scripps.edu/landing_page.php?pgcontent=mainPage) in another popular tool for processing of LC-MS and GC-MS-based data as well as raw metabolomics data storage sharing.

MetaboNexus is an interactive metabolomics data analysis platform that integrates preprocessing of raw peak data with in-depth statistical analysis and metabolite identity searching [83]. *MetaboNexus* allows users to conduct PCA, PLS-DA, other multi- and uni-variate analyses (e.g., t -test, ANOVA, Mann-Whitney U test, and Kruskal-Wallis test), among others.

MASSyPup64 (http://www.lababi.bioprocess.org/index.php/14-sample-data-articles/78-massypup) is a complete work flow that is Linux-based and allows data preprocessing on packages such as OpenMS/TOPPAS framework, the Trans-Proteomic-Pipeline (*TPP*) programs, the ProteoWizard tools, X!Tandem, Comet, and SpiderMass alongside statistical computing language R with packages for MS data analyses, such as XCMS/metaXCMS and *MetabR* [84], and is a standalone tool.

MetaboLyzer (https://sites.google.com/a/georgetown.edu/fornace-lab-informatics/home/metabolyzer) is a statistical analysis workflow that handles postprocessed LC-MS-based metabolomic datasets and performs a wide range of statistical tests, analytical procedures, and methodologies, in addition to rapid feature identification and biologically relevant analysis via incorporation of four major small molecule databases: *KEGG*, *HMDB*, *Lipid Maps*, and *BioCyc* [85].

Developed and maintained by the French Bioinformatics Institute (IFB) and the French Metabolomics and Fluxomics Infrastructure (MetaboHUB), *Workflow4Metabolomics* (W4M) (http://workflow4metabolomics.org/) helps users develop complete workflows for data preprocessing, statistical analysis, and annotation [86]. W4M is an open-source and collaborative online platform for computational metabolomics that is built upon Galaxy and is downloadable as a virtual machine for local installation.

Haystack (http://binf-app.host.ualr.edu/haystack/) is another online tool for rapid processing and analysis of LC-MS-based metabolomics data and is designed to visualize, parse, filter, and extract significant features [87].

MassCascade is an open-source library for LC-MS data processing where the available functions have been encapsulated in a plug-in for the workflow environment Konstanz Information Miner (KNIME) (https://www.knime.org/) [88], allowing combined use with other statistical workflow nodes [89].

3.8 Metadata, Data Handling, and Archiving

Metabolonote (http://metabolonote.kazusa.or.jp/) takes the innovative approach of using a Wiki -style interface to facilitate community-based metadata annotation of metabolomics datasets from the literature, thus delocalizing the burden of work [90]. *ChemDataExtractor* (http://chemdataextractor.org/) is a user-friendly webserver for reliable and automated extraction of chemical information from scientific literature, when provided as a document or pasted onto the server [91]. Moreover, the webserver can extract chemical information from journal articles, theses, patents, resolve the chemical names, abbreviations, and identifiers and extract experimental properties and spectroscopic information which can be readily used.

Gaggle (http://gaggle.systemsbiology.net/docs/) is an open-source software system for integrating bioinformatics software and data sources.

Taverna (http://www.taverna.org.uk/) is an open-source tool for designing and executing bioinformatics workflow allowing metadata management [92]. Similarly, *KNIME* is an open-source data analytics, reporting and integration platform.

Further, it is usually recommended to store and back-up the data on the user's hard (Storage) drives, local network in the academic (or, research) institution, and now in Cloud [93]. Nonetheless commercial Cloud-based storage resources such as those offered from Amazon and Google are also available for purchase.

3.9 Data Integration

Once the single-cell and single-cell-type metabolomics data are acquired, data sets available from other parallel –omics data generation technologies such as genomics, transcriptomics, and proteomics may be available for integration for aiding in robust biological interpretation (*see* **Note 9, 10**). For integrating the metabolomics datasets with other omics such as transcriptomics and proteomics datasets several on line web servers such as *Paintomics* 3 (http://bioinfo.cipf.es/paintomics/) [94], *3Omics* (http://3omics.cmdm.tw/) [95], *MassTRIX* [96] (http://masstrix3.helmholtz-muenchen.de/masstrix3/), *IMPaLa* (http://impala.molgen.mpg.de/wsdoc) [97] and software packages such as *mixOmics* (http://mixomics.org/), and *InCroMAP* (http://www.ra.cs.uni-tuebingen.de/software/InCroMAP/) [98] are available. Moreomics data integration tools and approaches are discussed elsewhere [99, 100]. However, efforts from single-cell multiomics [101] to understanding of cellular functions are an ongoing and continuous effort.

4 Notes

1. A common difficulty encountered in single-cell and single-cell-type metabolomics experiments is the lack of methods for absolute quantification of metabolites as well as that for normalization of data. For instance, in single-cell-type studies such as those for stomatal guard cells and trichomes, cell counts (in millions) or fresh weight of cell and total soluble protein content are used to normalize the data. However, the former approach is far from accurate, while for the later approach it can be said that protein turnover is not consistent with metabolite accumulation/depletion rates in a cell at a given time (i.e., steady-state assumption) or condition. The community also needs cell-specific normalization constants to deal with technical issues and variability by spike -in controls, internal standards, etc., or other innovative approaches.

2. The heterogeneity at the single-cell level is huge. Even single cells produced in the same batch on a single Petri-dish (animal cell line), single bacterial colony, or a shake-flask (in vitro plant suspension cultures) are known to show genetic aberrations, epigenetic changes, and other –omics scale changes which surely would reflect on their metabolic status in the captured metabolomics data. Largely due to the microenvironment to which each single cell is exposed to, beyond human control, it is difficult to overcome the associated changes.
3. Most single-cell and single-cell-type studies rely on artificial media to either grow, fix, mount, or extract the cells before analysis by a MS- or spectroscopy platforms. Now these experimental set up are far from 'cells in their native state' or as a part of a tissue within a whole organism, and hence, it is difficult to account for these confounding experimental factors, no matter whatever control one uses. Even for microfluidics based devices used upstream of the MS platforms, one cannot disregard the artificial buffers, flows, temperatures, probes, and (physical) pressures exerted on the single cell(s).
4. Purity of single-cell types is another challenge that needs to be addressed by the biology community that is interested in specific cell-types. No matter what protocol is used to obtain the cell-types, be it fluorescence-activated cell-sorting (FACS)-, microfluidic-, filtration-, centrifugation-, gradient-, or enzymatic-digestion-based, the preparation is always 90–99% pure and is only an enrichment. There are no realistic estimates as to how much contribution of metabolites comes from the other cell types that are adulterating, cellular debris and artefacts, and organelles among others (i.e., leachates from contaminating cell -types).
5. Another challenge is to obtain more specificity, lower noise, and higher signal from single-cell and single-cell-type metabolomics studies. Often, only the high abundance handful number of metabolites are captured. Unlike, the single-cell sequencing community, the deconvolution of the cell -types remains challenging just based on the metabolomics datasets- still awaiting the development of advanced reliable tools.
6. Obtaining spatial resolution for single-cell metabolomes is primarily limited by the size of the probe (laser beam or aspirator), and downstream challenges in image processing and data visualization. Metabolite "profiling" (i.e., capturing only a handful of chemical classes of metabolites) of single-cell types in complex human and plant tissues are feasible, but to obtain metabolomics-scale data remains challenging. To fix the cells or to subject them for analysis in live conditions (real time) remains challenging. However, recently it was shown in plant

cells that direct metabolomics of cells by live single-cell MS is doable in an efficient and informative manner, with suitable modifications and sophisticated instrumentation [102]. In cases, where discriminating cell lineages in a mixture of cells is difficult, clustering- or signature-based approaches would be quite useful. Tools that need to handle time-course and spatial data need to be developed for single-cell-specific applications. With newer techniques such as single-cell encapsulator, single-cell analysis chips, barcode chips and M3 emitters being continually developed, in future these tools would help acquire robust and high-throughput in vivo metabolomics data.

7. Tools and algorithms that take into account the intrinsic features of single-cell metabolomics experiments is lacking. Tools for cell population analysis using either Bayesian or beta-Poisson, machine learning (ML), deep learning (DL), dimensionality reduction tools (PCA, ICA sPCA), unsupervised clustering algorithms (density clustering, hierarchical clustering, k-means clustering) models already popular in RNAseq/ transcriptomics/ genomics domain are either missing currently for the metabolomics community or are in the developmental phases. Tools that can detect more complex and subtle changes than a mean shift are lacking altogether. Tools to correct for correction of confounding sources of variation, for normalization of abundance, quantification, cell-specific biases, detection of highly variable and significantly changing metabolites are in still in development.
8. Data integration from multiple MS-based platforms, that is, GC-MS, LC-MS, CE-MS, LAESI-MS, and/or MALDI-TOF-MS-based approaches within the single-cell metabolomics datasets, and combining them with the other omics such as transcriptomics, genomics, and proteomics remains another humongous challenge too.
9. Data acquisition and storage can be highly challenging for the spatial data. Cloud-based frameworks are necessary to serve the community in terms of saving analytics space and efforts in data storage, facilitating handling, and seamless, reproducible, and computational workflow implementation.
10. Data archiving efforts in the metabolomics community is fast catching up, unlike standards that are well-established in the genomics and proteomics communities, and is quickly improving. Even though archiving resources such as *MetaboLights* [25] and the ISA-TAB data format [103] options are available to facilitate metadata archiving/sharing, and *Metabolomics Workbench* and *GNPS-Massive* are available, only a handful of single-cell and single-cell-type metabolomics datasets are currently available in these databases. There is a great need for

development and availability of "gold standard" spatiotemporally resolved single-cell and single-cell-type metabolomics data sets for the metabolomics community to help compare and explore to address various analytical and informatics challenges.

5 Conclusions

Multitude of challenges galore, as far as the harmonized gold standard datasets, tools, and resources available for the single-cell and single-cell-type metabolomics tools and resources is concerned, a research area that is stll in it's developmental phase, and especially when compared to its sequencing (RNAseq and genomics), and sister mass-spectrometry (i.e., proteomics) counterparts [6, 68, 83, 104, 105]. In addition, the single-cell proteomics (SCP) community is dealing with similar challenges, and hence, there is a need for both the communities to come forward, forge a merger, and move in a unified direction to tackle the challenges in MS-based –single-cell –omics studies. For the time being, one has to use the existent tools that are currently being interchangeably used for whole organism or large-scale (bulk) plant, animal, and microbial metabolomics efforts. Although, currently limited by the available tools and techniques, understanding of metabolomic diversity at a single-cell resolution will promote understanding of how individual parts of a cell are integrated in time and space to form dynamic cellular processes as well as in uncovering of the relationship between cellular heterogeneity and metabolic regulation, which may eventually result in an understanding of disease statuses to drive therapeutic interventions in the future. As we clearly see that "tools" for single-cell and cell-type metabolomics are a "big black box" which needs to be opened and addressed by the growing single-cell and single-cell-type metabolomics research community.

Acknowledgements

The author thanks numerous pioneers in mass-spectrometry based metabolomics and single-cell and single cell-type -omics research, the developers and inventors of software tools, resources, and databases in metabolomics research who have inspired this compilation. The author also apologizes to the creators of numerous tools, resources, and analytical innovations that could not find a place in this chapter due to limitation in space or inadvertently.

References

1. Misra BB, Assmann SM, Chen S (2014) Plant single-cell and single-cell-type metabolomics. Trends Plant Sci 19(10):637–646
2. Zenobi R (2013) Single-cell metabolomics: analytical and biological perspectives. Science 342(6163):1243259
3. Uhlén M, Fagerberg L, Hallström BM, Lindskog C, Oksvold P, Mardinoglu A, Sivertsson Å, Kampf C, Sjöstedt E, Asplund A, Olsson I (2015) Tissue-based map of the human proteome. Science 347 (6220):1260419
4. Melé M, Ferreira PG, Reverter F, DeLuca DS, Monlong J, Sammeth M, Young TR, Goldmann JM, Pervouchine DD, Sullivan TJ, Johnson R (2015) The human transcriptome across tissues and individuals. Science 348 (6235):660–665
5. Bock C, Farlik M, Sheffield NC (2016) Multi-omics of single cells: strategies and applications. Trends Biotechnol 34(8):605–608
6. Shapiro E, Biezuner T, Linnarsson S (2013) Single-cell sequencing-based technologies will revolutionize whole-organism science. Nat Rev Genet 14(9):618–630
7. Misra BB, der Hooft JJ (2016) Updates in metabolomics tools and resources: 2014–2015. Electrophoresis 37(1):86–110
8. Misra BB (2016) Quick tips to perform a metabolomics study (No. e2002v1). Peer J Preprints 4:e2002v1
9. Misra BB (2018) New tools and resources in metabolomics: 2016–2017. Electrophoresis 39(7):909–923
10. Misra BB, Fahrmann JF, Grapov D (2017) Review of emerging metabolomic tools and resources: 2015–2016. Electrophoresis 38 (18):2257–2274
11. Misra BB, Langefeld CD, Olivier M, Cox LA (2018) Integrated omics: tools, advances, and future approaches. J Mol Endocrinol pii:JME-18-0055
12. Henry VJ, Bandrowski AE, Pepin AS, Gonzalez BJ, Desfeux A (2014) OMICtools: an informative directory for multi-omic data analysis. Database 2014:bau069
13. Lo SJ, Yao DJ (2015) Get to understand more from single-cells: current studies of microfluidic-based techniques for single-cell analysis. Int J Mol Sci 16(8):16763–16777
14. Link H, Fuhrer T, Gerosa L, Zamboni N, Sauer U (2015) Real-time metabolome profiling of the metabolic switch between starvation and growth. Nat Methods 12 (11):1091–1097
15. Broadhurst D, Goodacre R, Reinke SN, Kuligowski J, Wilson ID, Lewis MR, Dunn WB (2018) Guidelines and considerations for the use of system suitability and quality control samples in mass spectrometry assays applied in untargeted clinical metabolomic studies. Metabolomics 14(6):72
16. Lisec J, Schauer N, Kopka J, Willmitzer L, Fernie AR (2006) Gas chromatography mass spectrometry-based metabolite profiling in plants. Nat Protoc 1(1):387–396
17. Zhu ZJ, Schultz AW, Wang J, Johnson CH, Yannone SM, Patti GJ, Siuzdak G (2013) Liquid chromatography quadrupole time-of-flight mass spectrometry characterization of metabolites guided by the METLIN database. Nat Protoc 8(3):451–460
18. Noack S, Wiechert W (2014) Quantitative metabolomics: a phantom? Trends Biotechnol 32(5):238–244
19. Palmer A, Trede D, Alexandrov T (2016) Where imaging mass spectrometry stands: here are the numbers. Metabolomics 12 (6):1–3
20. Bartels B, Svatoš A (2015) Spatially resolved in vivo plant metabolomics by laser ablation-based mass spectrometry imaging (MSI) techniques: LDI-MSI and LAESI. Front Plant Sci 6:471
21. Jacobson RS, Thurston RL, Shrestha B, Vertes A (2015) In situ analysis of small populations of adherent mammalian cells using laser ablation electrospray ionization mass spectrometry in transmission geometry. Anal Chem 87(24):12130–12136
22. Paglia G, Williams JP, Menikarachchi L, Thompson JW, Tyldesley-Worster R, Halldórsson S, Rolfsson O, Moseley A, Grant D, Langridge J, Palsson BO (2014) Ion mobility derived collision cross sections to support metabolomics applications. Anal Chem 86(8):3985–3993
23. Perez-Riverol Y, Gatto L, Wang R, Sachsenberg T, Uszkoreit J, Leprevost F, Fufezan C, Ternent T, Eglen SJ, Katz DS, Pollard TJ (2016) Ten simple rules for taking advantage of git and GitHub. bioRxiv 048744
24. Boekel J, Chilton JM, Cooke IR, Horvatovich PL, Jagtap PD, Käll L, Lehtiö J, Lukasse P, Moerland PD, Griffin TJ (2015) Multi-omic data analysis using galaxy. Nat Biotechnol 33 (2):137–139
25. Haug K, Salek RM, Conesa P, Hastings J, de Matos P, Rijnbeek M, Mahendraker T,

Williams M, Neumann S, Rocca-Serra P, Maguire E (2012) MetaboLights—an open-access general-purpose repository for metabolomics studies and associated meta-data. Nucleic Acids Res 41(D1):D781–D786

26. Sud M, Fahy E, Cotter D, Azam K, Vadivelu I, Burant C, Edison A, Fiehn O, Higashi R, Nair KS, Sumner S (2015) Metabolomics workbench: an international repository for metabolomics data and metadata, metabolite standards, protocols, tutorials and training, and analysis tools. Nucleic Acids Res 44 (D1):D463–D470

27. Wang M, Carver JJ, Phelan VV, Sanchez LM, Garg N, Peng Y, Nguyen DD, Watrous J, Kapono CA, Luzzatto-Knaan T, Porto C (2016) Sharing and community curation of mass spectrometry data with Global Natural Products Social Molecular Networking. Nat Biotechnol 34(8):828–837

28. Wang R, Perez-Riverol Y, Hermjakob H, Vizcaíno JA (2015) Open source libraries and frameworks for biological data visualisation: a guide for developers. Proteomics 15 (8):1356–1374

29. Weiskirchen R, Weiskirchen S, Kim P, Winkler R (2019) Software solutions for evaluation and visualization of laser ablation inductively coupled plasma mass spectrometry imaging (LA-ICP-MSI) data: a short overview. J Cheminform 11(1):16. https://doi.org/10.1186/s13321-019-0338-7

30. Wang D, Bodovitz S (2010) Single cell analysis: the new frontier in 'omics'. Trends Biotechnol 28(6):281–290

31. Vasilevsky N, Johnson T, Corday K et al (2012) Research resources: curating the new Eagle-I Discovery System. Database (Oxford) 2012:bar067

32. Wilkinson MD, Dumontier M, Aalbersberg IJ, Appleton G, Axton M, Baak A, Blomberg N, Boiten JW, da Silva Santos LB, Bourne PE, Bouwman J (2016) The FAIR guiding principles for scientific data management and stewardship. Sci Data 3:160018

33. Holman JD, Tabb DL, Mallick P (2014) Employing ProteoWizard to convert raw mass spectrometry data. Curr Protoc Bioinformatics 46:13.24.1–13.24.9

34. Wenig P, Odermatt J (2010) OpenChrom: a cross-platform open source software for the mass spectrometric analysis of chromatographic data. BMC Bioinformatics 11 (1):1

35. Strohalm M, Kavan D, Novak P, Volny M, Havlicek V (2010) mMass 3: a cross-platform software environment for precise analysis of mass spectrometric data. Anal Chem 82 (11):4648–4651

36. Pluskal T, Castillo S, Villar-Briones A, Orešič M (2010) MZmine 2: modular framework for processing, visualizing, and analyzing mass spectrometry-based molecular profile data. BMC Bioinformatics 11(1):1

37. Tautenhahn R, Cho K, Uritboonthai W, Zhu Z, Patti GJ, Siuzdak G (2012) An accelerated workflow for untargeted metabolomics using the METLIN database. Nat Biotechnol 30(9):826–828

38. Tautenhahn R, Patti GJ, Rinehart D, Siuzdak G (2012) XCMS online: a web-based platform to process untargeted metabolomic data. Anal Chem 84(11):5035–5039

39. Gowda H, Ivanisevic J, Johnson CH, Kurczy ME, Benton HP, Rinehart D, Nguyen T, Ray J, Kuehl J, Arevalo B, Westenskow PD (2014) Interactive XCMS online: simplifying advanced metabolomic data processing and subsequent statistical analyses. Anal Chem 86(14):6931–6939

40. Smith CA, Want EJ, O'Maille G, Abagyan R, Siuzdak G (2006) XCMS: processing mass spectrometry data for metabolite profiling using nonlinear peak alignment, matching, and identification. Anal Chem 78 (3):779–787

41. Tanaka S, Fujita Y, Parry HE, Yoshizawa AC, Morimoto K, Murase M, Yamada Y, Yao J, Utsunomiya SI, Kajihara S, Fukuda M (2014) Mass++: a visualization and analysis tool for mass spectrometry. J Proteome Res 13(8):3846–3853

42. Tsugawa H, Cajka T, Kind T, Ma Y, Higgins B, Ikeda K, Kanazawa M, VanderGheynst J, Fiehn O, Arita M (2015) MS-DIAL: data-independent MS/MS deconvolution for comprehensive metabolome analysis. Nat Methods 12(6):523–526

43. Clasquin MF, Melamud E, Rabinowitz JD (2012) LC-MS data processing with MAVEN: a metabolomic analysis and visualization engine. Curr Protoc Bioinformatics 4:14–11

44. Davidson RL, Weber RJ, Liu H, Sharma-Oates A, Viant MR (2016) Galaxy-M: a Galaxy workflow for processing and analyzing direct infusion and liquid chromatography mass spectrometry-based metabolomics data. GigaScience 5(1):1

45. Davies T (1998) The new automated mass spectrometry deconvolution and identification system (AMDIS). Spectrosc Eur 10 (3):24–27

46. Sturm M, Bertsch A, Gröpl C, Hildebrandt A, Hussong R, Lange E, Pfeifer N, Schulz-Trieglaff O, Zerck A, Reinert K, Kohlbacher O (2008) OpenMS—an open-source software framework for mass spectrometry. BMC Bioinformatics 9(1):163
47. Ma Y, Kind T, Yang D, Leon C, Fiehn O (2014) MS2Analyzer: a software for small molecule substructure annotations from accurate tandem mass spectra. Anal Chem 86 (21):10724–10731
48. Kind T, Liu KH, Lee DY, DeFelice B, Meissen JK, Fiehn O (2013) LipidBlast in silico tandem mass spectrometry database for lipid identification. Nat Methods 10(8):755–758
49. Lommen A (2012) Data (pre-) processing of nominal and accurate mass LC-MS or GC-MS data using MetAlign. Plant Metabolomics 860:229–253
50. Jaitly N, Mayampurath A, Littlefield K, Adkins JN, Anderson GA, Smith RD (2009) Decon2LS: an open-source software package for automated processing and visualization of high resolution mass spectrometry data. BMC Bioinformatics 10(1):1
51. Parry RM, Galhena AS, Gamage CM, Bennett RV, Wang MD, Fernández FM (2013) omniSpect: an open MATLAB-based tool for visualization and analysis of matrix-assisted laser desorption/ionization and desorption electrospray ionization mass spectrometry images. J Am Soc Mass Spectrom 24 (4):646–649
52. O'Connor PB (2002) Boston University data analysis (BUDA). Boston University, Boston, MA. http://www.bumc.bu.edu/ftms/buda
53. Kopka J, Schauer N, Krueger S, Birkemeyer C, Usadel B, Bergmüller E, Dörmann P, Weckwerth W, Gibon Y, Stitt M, Willmitzer L (2005) GMD@ CSB. DB: the Golm metabolome database. Bioinformatics 21(8):1635–1638
54. Wishart DS, Jewison T, Guo AC, Wilson M, Knox C, Liu Y, Djoumbou Y, Mandal R, Aziat F, Dong E, Bouatra S (2012) HMDB 3.0—the human metabolome database in 2013. Nucleic Acids Res 41(Database issue): D801–D807
55. Kanehisa M, Goto S, Sato Y, Kawashima M, Furumichi M, Tanabe M (2014) Data, information, knowledge and principle: back to metabolism in KEGG. Nucleic Acids Res 42 (D1):D199–D205
56. Horai H, Arita M, Kanaya S, Nihei Y, Ikeda T, Suwa K, Ojima Y, Tanaka K, Tanaka S, Aoshima K, Oda Y (2010) MassBank: a public repository for sharing mass spectral data for life sciences. J Mass Spectrom 45(7):703–714
57. Mistrik R, Lutisan J, Huang Y, Suchy M, Wang J, Raab M (2013) mzCloud: a key conceptual shift to understand 'Who's Who'in untargeted metabolomics. In Metabolomics society 2013 conference, Glasgow, July 2013, pp. 1–4
58. Sawada Y, Nakabayashi R, Yamada Y, Suzuki M, Sato M, Sakata A, Akiyama K, Sakurai T, Matsuda F, Aoki T, Hirai MY (2012) RIKEN tandem mass spectral database (ReSpect) for phytochemicals: a plant-specific MS/MS-based data resource and database. Phytochemistry 82:38–45
59. Afendi FM, Okada T, Yamazaki M, Hirai-Morita A, Nakamura Y, Nakamura K, Ikeda S, Takahashi H, Altaf-Ul-Amin M, Darusman LK, Saito K (2012) KNApSAcK family databases: integrated metabolite–plant species databases for multifaceted plant research. Plant Cell Physiol 53(2):e1–e1
60. Jeffryes JG, Colastani RL, Elbadawi-Sidhu M, Kind T, Niehaus TD, Broadbelt LJ, Hanson AD, Fiehn O, Tyo KE, Henry CS (2015) MINEs: open access databases of computationally predicted enzyme promiscuity products for untargeted metabolomics. J Cheminform 7(1):1
61. Li L, Li R, Zhou J, Zuniga A, Stanislaus AE, Wu Y, Huan T, Zheng J, Shi Y, Wishart DS, Lin G (2013) MyCompoundID: using an evidence-based metabolome library for metabolite identification. Anal Chem 85 (6):3401–3408
62. Allen F, Pon A, Wilson M, Greiner R, Wishart D (2014) CFM-ID: a web server for annotation, spectrum prediction and metabolite identification from tandem mass spectra. Nucleic Acids Res 42(W1):W94–W99
63. Chawade A, Alexandersson E, Levander F (2014) Normalyzer: a tool for rapid evaluation of normalization methods for omics data sets. J Proteome Res 13(6):3114–3120
64. Smyth GK, Speed T (2003) Normalization of cDNA microarray data. Methods 31 (4):265–273
65. Husson, F., Josse, J., Le, S., Mazet, J. and Husson, M.F., 2016. Package 'FactoMineR'
66. Grapov D (2014) DeviumWeb: dynamic multivariate data analysis and visualization platform. doi:https://doi.org/10.5281/zenodo.17459. https://github.com/dgrapov/DeviumWeb
67. Chong J, Soufan O, Li C, Caraus I, Li S, Bourque G, Wishart DS, Xia J (2018) MetaboAnalyst 4.0: towards more transparent and

integrative metabolomics analysis. Nucleic Acids Res 46(W1):W486–W494

68. López-Ibáñez J, Pazos F, Chagoyen M (2016) MBROLE 2.0—functional enrichment of chemical compounds. Nucleic Acids Res 44(W1):W201–W204
69. Batchelor C, Brenninkmeijer C, Chichester C, Davies M, Digles D, Dunlop I, Evelo CT, Gaulton A, Goble C, Gray AJG, Groth P, Harland L, Karapetyan K, Loizou A, Overington JP, Pettifer S, Steele J, Stevens R, Tkachenko V, Waagmeester A, Williams A, Willighagen EL (2014) Scientific lenses to support multiple views over linked chemistry data. In The semantic Web – ISWC 2014. Lect Notes Comput Sci 8796:98–113
70. Jewison T, Su Y, Disfany FM, Liang Y, Knox C, Maciejewski A, Poelzer J, Huynh J, Zhou Y, Arndt D, Djoumbou Y (2013) SMPDB 2.0: big improvements to the small molecule pathway database. Nucleic Acids Res 42(Database issue):D478–D484
71. Caspi R, Billington R, Foerster H, Fulcher CA, Keseler I, Kothari A, Krummenacker M, Latendresse M, Mueller LA, Ong Q, Paley S (2016) BioCyc: online resource for genome and metabolic pathway analysis. FASEB J 30 (1 Suppl):lb192
72. Kutmon M, van Iersel MP, Bohler A, Kelder T, Nunes N, Pico AR, Evelo CT (2015) PathVisio 3: an extendable pathway analysis toolbox. PLoS Comput Biol 11(2): e1004085
73. Fitzpatrick MA, McGrath CM, Young SP (2014) Pathomx: an interactive workflow-based tool for the analysis of metabolomic data. BMC Bioinformatics 15(1):1
74. Kaever A, Landesfeind M, Feussner K, Mosblech A, Heilmann I, Morgenstern B, Feussner I, Meinicke P (2015) MarVis-Pathway: integrative and exploratory pathway analysis of non-targeted metabolomics data. Metabolomics 11(3):764–777
75. Szklarczyk D, Santos A, von Mering C, Jensen LJ, Bork P, Kuhn M (2015) STITCH 5: augmenting protein–chemical interaction networks with tissue and affinity data. Nucleic Acids Res 44:D380–D384
76. Cottret L, Wildridge D, Vinson F, Barrett MP, Charles H, Sagot MF, Jourdan F (2010) MetExplore: a web server to link metabolomic experiments and genome-scale metabolic networks. Nucleic Acids Res 38(Suppl 2): W132–W137
77. Narang P, Khan S, Hemrom AJ, Lynn AM (2014) MetaNET-a web-accessible interactive platform for biological metabolic network analysis. BMC Syst Biol 8(1):1
78. Joshi-Tope G, Gillespie M, Vastrik I, D'Eustachio P, Schmidt E, de Bono B, Jassal B, Gopinath GR, Wu GR, Matthews L, Lewis S (2005) Reactome: a knowledgebase of biological pathways. Nucleic Acids Res 33 (Suppl 1):D428–D432
79. Shannon P, Markiel A, Ozier O, Baliga NS, Wang JT, Ramage D, Amin N, Schwikowski B, Ideker T (2003) Cytoscape: a software environment for integrated models of biomolecular interaction networks. Genome Res 13(11):2498–2504
80. Shannon PT, Reiss DJ, Bonneau R, Baliga NS (2006) The Gaggle: an open-source software system for integrating bioinformatics software and data sources. BMC Bioinformatics 7(1):1
81. Pathan M, Keerthikumar S, Ang CS, Gangoda L, Quek CY, Williamson NA, Mouradov D, Sieber OM, Simpson RJ, Salim A, Bacic A (2015) FunRich: an open access standalone functional enrichment and interaction network analysis tool. Proteomics 15(15):2597–2601
82. Xia J, Wishart DS (2016) Using MetaboAnalyst 3.0 for comprehensive metabolomics data analysis. Curr Protoc Bioinformatics 55:14.10.1–14.10.91
83. Huang SM, Toh W, Benke PI, Tan CS, Ong CN (2014) MetaboNexus: an interactive platform for integrated metabolomics analysis. Metabolomics 10(6):1084–1093
84. Winkler R (2015) An evolving computational platform for biological mass spectrometry: workflows, statistics and data mining with MASSyPup64. PeerJ 3:e1401
85. Mak TD, Laiakis EC, Goudarzi M, Fornace AJ Jr (2013) Metabolyzer: a novel statistical workflow for analyzing postprocessed LC–MS metabolomics data. Anal Chem 86 (1):506–513
86. Giacomoni F, Le Corguillé G, Monsoor M, Landi M, Pericard P, Pétéra M, Duperier C, Tremblay-Franco M, Martin JF, Jacob D, Goulitquer S (2015) Workflow4Metabolomics: a collaborative research infrastructure for computational metabolomics. Bioinformatics 31(9):1493–1495
87. Grace SC, Embry S, Luo H (2014) Haystack, a web-based tool for metabolomics research. BMC Bioinformatics 15(11):1
88. Berthold MR, Cebron N, Dill F, Gabriel TR, Kötter T, Meinl T, Ohl P, Thiel K, Wiswedel B (2009) KNIME-the Konstanz information miner: version 2.0 and beyond. AcM SIGKDD Explor Newsl 11(1):26–31

89. Beisken S, Earll M, Portwood D, Seymour M, Steinbeck C (2014) MassCascade: visual programming for LC-MS data processing in metabolomics. Mol Inform 33(4):307–310
90. Ara T, Enomoto M, Arita M, Ikeda C, Kera K, Yamada M, Nishioka T, Ikeda T, Nihei Y, Shibata D, Kanaya S (2015) Metabolonote: a wiki-based database for managing hierarchical metadata of metabolome analyses. Front Bioeng Biotechnol 3:38
91. Swain MC, Cole JM (2016) ChemDataExtractor: a toolkit for automated extraction of chemical information from the scientific literature. J Chem Inf Model 56(10):1894–1904
92. Wolstencroft K, Haines R, Fellows D, Williams A, Withers D, Owen S, Soiland-Reyes S, Dunlop I, Nenadic A, Fisher P, Bhagat J (2013) The Taverna workflow suite: designing and executing workflows of Web Services on the desktop, web or in the cloud. Nucleic Acids Res 41(Web Server issue): W557–W561
93. Warth B, Levin N, Rinehart D, Teijaro J, Benton HP, Siuzdak G (2017) Metabolizing data in the cloud. Trends Biotechnol 35 (6):481–483
94. García-Alcalde F, García-López F, Dopazo J, Conesa A (2011) Paintomics: a web based tool for the joint visualization of transcriptomics and metabolomics data. Bioinformatics 27(1):137–139
95. Kuo TC, Tian TF, Tseng YJ (2013) 3Omics: a web-based systems biology tool for analysis, integration and visualization of human transcriptomic, proteomic and metabolomic data. BMC Syst Biol 7(1):1
96. Wägele B, Witting M, Schmitt-Kopplin P, Suhre K (2012) MassTRIX reloaded: combined analysis and visualization of transcriptome and metabolome data. PLoS One 7(7): e39860
97. Kamburov A, Cavill R, Ebbels TM, Herwig R, Keun HC (2011) Integrated pathway-level analysis of transcriptomics and metabolomics data with IMPaLA. Bioinformatics 27 (20):2917–2918
98. Eichner J, Rosenbaum L, Wrzodek C, Häring HU, Zell A, Lehmann R (2014) Integrated enrichment analysis and pathway-centered visualization of metabolomics, proteomics, transcriptomics, and genomics data by using the InCroMAP software. J Chromatogr B 966:77–82
99. Wanichthanarak K, Fahrmann JF, Grapov D (2015) Genomic, proteomic, and metabolomic data integration strategies. Biomarker Insights 10(Suppl 4):1
100. Misra BB, Mohapatra S (2019 Jan) Tools and resources for metabolomics research community: a 2017–2018 update. Electrophoresis 40 (2):227–246
101. Macaulay IC, Ponting CP, Voet T (2017) Single-cell multiomics: multiple measurements from single cells. Trends Genet 33 (2):155–168
102. Fujii T, Matsuda S, Tejedor ML, Esaki T, Sakane I, Mizuno H, Tsuyama N, Masujima T (2015) Direct metabolomics for plant cells by live single-cell mass spectrometry. Nat Protoc 10(9):1445–1456
103. Rocca-Serra P, Brandizi M, Maguire E, Sklyar N, Taylor C, Begley K, Field D, Harris S, Hide W, Hofmann O, Neumann S (2010) ISA software suite: supporting standards-compliant experimental annotation and enabling curation at the community level. Bioinformatics 26(18):2354–2356
104. Le Cao KA, Gonzalez I, Dejean S, Rohart F, Gautier B, Monget P, Coquery J, Yao F, Liquet B (2015) Package 'mixOmics'
105. Onjiko RM, Moody SA, Nemes P (2015) Single-cell mass spectrometry reveals small molecules that affect cell fates in the 16-cell embryo. Proc Natl Acad Sci 112 (21):6545–6550

Chapter 16

Ten Major Future Challenges in Single-Cell Metabolomics

Bindesh Shrestha

Abstract

Information on cellular metabolism at the single-cell level can unravel countless biochemical process providing invaluable biomedical insight. Single-cell analysis field is at the very early stage at this moment, and all the work done so far are proof-of-principle work by early-stage researchers. In this chapter, I have outlined ten fundamental issues that are required for the development of robust single-cell metabolomics platform using mass spectrometry (MS).

Key words Single-cell analysis, Single-cell omic, Single-cell mass spectrometry, Single-cell metabolomics, Single-cell future directions

1 Introduction

Single-cell metabolite analysis can provide information on endogenous or exogenous small molecules without inferences from other neighboring cells. Single-cell analysis can provide critical biochemical information unique to the cell's phenotype. Methods for single-cell metabolism analysis using MS are still in their infancy. Several development and challenges remain to move the technology from the hand of expert analytical scientists to practicing biomedical scientists.

Some of the fundamental challenges regarding single-cell analysis by mass spectrometry are outlined here. Solving these challenges is going to be vital in developing the field.

1.1 Coverage of Metabolites

Many analytical tools do not have broad coverage of metabolites because of the lack of sensitivity and selectivity in the complex biological matrix, such as cells. Thus, different analytical platforms may be needed for complete single-cell metabolomics studies. Analysis of the metabolome has to account for the diversity of chemical makeup of metabolites and smaller analytical volume of single cells. Generally, there is an inverse relationship between molecular coverage and sample volume size. For example, ion

Bindesh Shrestha (ed.), *Single Cell Metabolism: Methods and Protocols*, Methods in Molecular Biology, vol. 2064,
https://doi.org/10.1007/978-1-4939-9831-9_16,

beam in secondary-ion MS (SIMS) can be routinely focused to around hundreds of nanometers, but limited to only a few selected molecules. Matrix-assisted laser desorption ionization (MALDI) has shown single-cell level pixel size but often detected metabolites are limited to the ones that are most abundant and easily undergo ionization. The sensitivity for certain class of molecule can be enhanced by selective derivitization. Another approach to increasing molecular coverage is to use a targeted approach, such as in mass cytometry, where hard-to-detect analytes are tagged and easily detected. Further instrumental and method development work will be needed to improve the coverage of metabolites.

1.2 Identification of Metabolites

Accurate identification of metabolites is key for any biomedical data interpretation. Metabolite identification is straightforward in the targeted analysis where instrument response is carefully established for each metabolite, such as reporter ions in the MS/MS spectra. However, metabolite identification, or more accurately annotation, is a major challenge in untargeted metabolomics for discovery. A number of potential metabolite candidates can be narrowed by using the accurate mass from any modern high-resolution mass spectrometer. Increased confidence in annotated metabolites candidate can be obtained by simultaneously acquiring the MS scan and MS/MS using data-dependent acquisition (DDA) and data-independent acquisition (DIA) MS techniques. Both DDA and DIA acquire a full MS scan. In DDA, precursor ions acquired in the MS scan are ranked, and selected subset, such as most abundant ions, are fragmented for MS/MS information. In DIA, all precursor ions in a predefined isolation window are fragmented. In theory, DDA offers cleaner fragmentation information on the limited set of molecules, while the DIA technique provides more extensive selection but more complex mixed fragmentation information. New development in DIA techniques such as SONAR utilizes a continuously scanning precursor window for fragmentation resulting in cleaner fragmentation information for positive correlation of precursor and fragment ions.

1.3 Sensitivity or Limit of Detection

One of the key performances of an analytical device such as mass spectrometers during the analysis of a volume-limited sample, such as a single cell, is the sensitivity. Higher sensitivity is prerequisite for detecting when analyte copy number is low. There is an inverse correlation between the sensitivity of molecules and sample volume size. The sensitivity of an analytical instrument can be defined as the International Union of Pure and Applied Chemistry (IUPAC) as "*The slope of the calibration curve. If the curve is in fact a 'curve', rather than a straight line, then of course sensitivity will be a function of analyte concentration or amount. If sensitivity is to be a unique performance characteristic, it must depend only on the chemical measurement process, not upon scale factors.*" For mass spectrometers, higher sensitivity means an increase in the ion count due to better ionization, ion transmission, and ion detection. In a typical assay, the limit of detection (LOD) is defined as the smallest amount of analyte that can be detected with an adequate signal-to-noise ratio such as 3. We can anticipate improved instrumentation capabilities, and refined sample preparation steps will help enhance the overall sensitivity needed for routine single-cell metabolomics analysis.

1.4 Quantitative Measurement

Measuring the amount of metabolites with quantitative precision from a complex and volume-limited biological matrix is very challenging. A limit of quantitation (LoQ) is often found at a higher concentration than the limit of detection (LoD) and never below it. LoQ is defined as the lowest concentration at which the analyte can be reliably repeatability detected with a predetermined acceptable level of precision and error not exceeding measurement at other concentrations. Added internal standards, such as stable isotope-labeled counterparts or endogenous molecules, can be used to obtain quantitative information during single-cell metabolomics. Improved quantitation data analysis tools would be able to correlate relationships between hundreds of biomolecules (in mass spectra) and in the context of hundreds of single-cell analysis.

1.5 Metabolic Flux

Metabolism is a dynamic process, while metabolomics can provide information on the metabolites in a biological sample at that time point, metabolomic flux aims to provide the rates of metabolic reactions within that biological system. The biggest challenge to study metabolic flux is our ability to quantify rapidly changing metabolome. Metabolome at the level of single cells is a perfect biological system to study flux. However, in order to study the metabolome flux, the analytical technique would need multiple analysis of same single cells, which is not possible with destructive analytical technique such as MS. Currently, the one way to solve this challenge is by utilizing synchronized cell cultures and inspecting how metabolic fluxes in single cells varies during the cell cycle.

1.6 Spatial Distribution

Cell-by-cell MS imaging can put population information in the context of cell micro and macro environment. Some techniques have been used to get single-cell level imaging. Secondary ion mass spectrometry (SIMS) has been predominantly used for single-cell imaging by mass spectrometry due to its submicron desorption profiles. However, SIMS cannot ionize most of the metabolites within the cell. There have been technological and methodological developments in desorption platforms such as MALDI to perform successful single-cell metabolite imaging. Some other emerging techniques that are utilized for single-cell imaging include laser ablation electrospray ionization (LAESI), nanopost arrays

(NAPA), and nanostructure-initiator mass spectrometry (NIMS). Currently, most of the single-cell MS imaging methods lack a robust sensitivity to image medium-to-low abundance metabolites at the cellular spatial resolution.

1.7 Multiomics

Metabolite analysis of single-cell offers a dynamic snapshot of the functional aspect of individual cells. However, analysis of just one type of molecules, such as metabolites, provide incomplete information at best. Cellular processes are governed by the complex interplay of molecules illustrated by its genome, epigenome, transcriptome, proteome, and metabolome. A holistic multiomics study of a single cell can only be obtained by comprehensive analytical workflow encompassing all type of molecules. Multiomics approaches can lead to our understanding of single-cell genotype and phenotype relationships. A comprehensive understanding of these cellular metabolism processes will require new analytical tools as well as new data analysis platforms to build a relationship between assay different types of molecules—data from routine single-cell RNA-seq combined with the single-cell metabolomics and proteomics to study system biology.

1.8 Sorting and Sampling

Single cells can be sorted and isolated from a heterogeneous mixture of cells based on intracellular and extracellular properties by several methods. Some of these methods are already used for routine single-cell biological assays such as flow cytometry. The other methods include arrays of micro-wells, also known as microrafts that can be used sort and isolate single cells. In this method, cells are plated on an array of single-cell-sized micro-wells in the same manner as standard cell culture. After plating, single cells settle into the individual micro-wells that can be isolated for later analysis. The second method is a lab-on-a-chip single-cell platform. Lab-on-a-chip devices are composed of small microchannels for manipulating cells in solution cells using fluidics and small forces. They are often made with soft lithography techniques and polydimethylsiloxane (PDMS) and is considered gentler to cells resulting in better cell viability after sorting. In the future, we can anticipate these cell-sorting strategies combined with finding a way to mass spectrometric analysis, perhaps capture, manipulation, growth, and analysis of a single cell in one device.

1.9 High-Throughput Analysis

High-throughput analysis of single cells, such as flow cytometry, is needed to confidently measure rare cells that have a critical impact on biomedical processes within the body, such as disease progression in cancer. Analysis of a large number of single cells is prerequisite to get a statistically significant detection. For embedded cells in the tissue, we cannot use the sampling method used for free-flowing cells such as flow cytometer. For these cells, some type of desorption ionization workflow, such as high-throughput MALDI, is needed.

1.10 Reproducibility

Reproducible single-cell metabolomics data, both intralaboratory (several samples in an extended period) and interlaboratory (different labs and instruments) are critical for confident biomedical studies. Single-cell metabolomics experiment will need a standardized, well-defined sample harvesting, storage, extraction, and preparation steps in pre-analytical step and careful quality control steps during the analytical run steps, such as the introduction of internal standards to account for any analytical variance and bias, and normalization to a reference during the post-analysis step. Each single-cell analysis is a unique measurement that cannot be technical repeated. Reproducibility or precision of single-cell metabolomics analysis is essential to discover and characterize outlier cells within a population such as drug resistance cells in cancer therapy.

2 Conclusions

In the future, we can expect to see the steady development of the methods and tools to address these challenges that will enable us to decipher the metabolic signature of complex and heterogeneous cell populations at single-cell resolution. Advances in mass spectrometer instrumental performance will be vital in advancing single-cell research. The critical mass spectrometer instrument parameters include an increase in sensitivity, higher mass resolution, cell-sampling methods, new ionization approaches, coupling with separation platforms such as capillary electrophoresis (CE) and ion mobility separation (IMS). In addition to instrumental development, informatics tools need to be developed to analyze the considerable volume of single-cell data and find statistically significant outcomes. Finally, development in standardized protocols in single-cell sample preparation will be required for translational biomedical research and clinical applications.

INDEX

Bindesh Shrestha (ed.), *Single Cell Metabolism: Methods and Protocols*, Methods in Molecular Biology, vol. 2064, https://doi.org/10.1007/978-1-4939-9831-9, © Springer Science+Business Media, LLC, part of Springer Nature 2020

FSC
www.fsc.org
MIX
Papier | Fördert gute Waldnutzung
FSC® C083411

Kolibri 360 GmbH
Ferdinand-Jühlke-Straße 7
99095 Erfurt, Deutschland
produktsicherheit@kolibri360.de